Taking a practical approach to clinical medicine, this series of smaller reference books is designed for the trainee physician, primary care physician, nurse practitioner and other general medical professionals to understand each topic covered. The coverage is comprehensive but concise and is designed to act as a primary reference tool for subjects across the field of medicine.

More information about this series at http://www.springer. com/series/13483

Abdul Qayyum Rana • Ali T. Ghouse
Raghav Govindarajan

# Neurophysiology in Clinical Practice

 Springer

Abdul Qayyum Rana
Parkinson's Clinic of Eastern
Toronto and Movement
Disorders Centre
Toronto, ON
Canada

Raghav Govindarajan
Department of Neurology
University of Missouri
Columbia, MO
USA

Ali T. Ghouse
McMaster University
Department of Medicine
Hamilton, ON
Canada

In Clinical Practice
ISSN 2199-6652                      ISSN 2199-6660    (electronic)
ISBN 978-3-319-39341-4        ISBN 978-3-319-39342-1    (eBook)
DOI 10.1007/978-3-319-39342-1

Library of Congress Control Number: 2016953206

Printed on acid-free paper

This Springer imprint is published by Springer Nature
The registered company is Springer International Publishing AG
The registered company address is Gewerbestrasse 11, 6330 Cham, Switzerland

*To our beloved families, colleagues, students, and, most important, our patients who have inspired us to complete this work.*

# Preface

Neurophysiology rotations can be challenging for beginners, and they may require several weeks to become familiar with various aspects of neurophysiology. Even after residency and fellowship training, the practitioner may feel they still don't have mastery of the field. This handbook is designed to provide a basic introduction to neurophysiology for practicing physicians, advanced care providers, fellows, and residents, as well as medical students doing their neurophysiology rotation. This guide is meant to help the reader to understand the basics of neurophysiology and its clinical applications, but it is not a comprehensive review.

The handbook is divided into two sections. Part I (Chaps. 1, 2, 3) describes electroencephalography and Part II (Chaps. 4, 5, 6, 7, 8, 9, 10, 11, 12, 13, 14, 15, 16, 17, 18, 19, 20, 21, 22, and 23) discusses electromyography and nerve conduction studies. All of the information presented in this manual has been reviewed for accuracy and practice; however, the authors/editors and publisher are not responsible for any errors, omissions, or consequences arising from the application of this information, and they make no expressed or implied warranty of the contents of this publication. Suggestions to improve this publication are welcome and should be directed to the authors.

Toronto, ON, Canada
Abdul Qayyum Rana, MD, FRCPC, FRCP (HON)

Hamilton, ON, Canada
Ali T. Ghouse, MD, FRCPC

Columbia, MO, USA
Raghav Govindarajan, MD, FISQua, FACSc, FCPP

# Acknowledgments

We are thankful to Dr. Amza Ali and Dr. Danita Jones for reviewing and providing useful input on individual sections of this book. We are also grateful to our students Usman Saeed, Mohammad Rana, Mohamed Saleh, and Danial Qureshi for their tireless support in completing this work. We appreciate the contributions of Evelyn Shifflett and Hussain Cader in creating the illustrations, and we appreciate others who supported us to complete this work. Last but not least, we thank Joanna Renwick, editor Springer (UK), for her guidance in putting this book together.

# Contents

# Part I
# Electroencephalography

# Chapter 1
## Basics of Electroencephalography (EEG)

### Physiology

Electroencephalography (EEG) is the noninvasive recording and measuring of the electrical activity of the cerebral cortex, with surface EEG performed via electrodes placed on the scalp. The recorded voltage activity is the net difference between ionic current flows within the brain, controlled by N-methyl-D-aspartate (NMDA) glutamate receptors becoming permeable to calcium ions. An EEG recording is a summation of numerous different frequencies. The depolarization and synchronous activation of many neurons generates epileptiform activity. The spike and wave activity seen in epilepsy is likely caused by cyclic depolarization and repolarization. The inhibiting feedback of the neurons results in the ultimate cessation of the epileptiform activity. The thalamus is considered to be the main site for the origin of cortical excitability. A minimum of 6 cm$^2$ of cortical synchronous activity is needed to create a recordable scalp potential on surface EEG.

This short review of the basic principles of EEG recording and interpretation, while by no means exhaustive, should assist the reader in becoming competent in requesting and interpreting an EEG report.

A.Q. Rana et al., *Neurophysiology in Clinical Practice*,
In Clinical Practice, DOI 10.1007/978-3-319-39342-1_1,
© Springer International Publishing Switzerland 2017

# EEG Recording

Biological signals consist of various different sequences. Filters are used to exclude frequencies that are less useful. EEG machines have a 60-Hz filter, which removes activity in the 60-Hz range and helps in eliminating artifacts from the line voltage that affect physiological recordings. The use of different montages, that is, different organizations of scalp electrodes, allows for different methodologies to be used in reading the EEG recording. With these montages, the same event will vary because of the different electrode pairings.

The following are the frequencies of different waves seen in a routine EEG recording:

Delta waves have a frequency of 1–3.99 Hz.
Theta waves have a frequency of 4–7.99 Hz.
Alpha waves have a frequency of 8–12.99 Hz.
Beta waves have a frequency above 13 Hz.

TABLE I.I  Electroencephalographic waves

| Wave | Frequency | Normal | Abnormal |
|------|-----------|--------|----------|
| Alpha | 8–12.99 Hz | Dominant rhythm during wakefulness. May be identified by its reactivity and suppression by eye opening. Maximally seen over occipital deviations | Diffuse alpha pattern seen in coma |
| Beta | above 13 Hz | Normal sleep rhythm in young children. May become more prominent in adults when changing from wakefulness to drowsiness | Amplitudes >50 seen with barbiturate and benzodiazepine use. Lower unilateral voltage may represent cerebral edema or dural fluid collections. Loss of beta activity implies cortical injury |

TABLE I.I (continued)

| Wave | Frequency | Normal | Abnormal |
|------|-----------|--------|----------|
| Delta | 1–3.99 Hz | Normal sleep rhythm. Delta waves may be induced by hyperventilation. Found in one-third of adults within the frontocentral regions during eye-closed wakefulness | Intermittent rhythmic delta activity with polymorphic delta activity in focal lesions |
| Theta | 4–7.99 Hz | Normal in sleep and drowsiness, may be present in posterior slow waves of youth | Temporal theta pattern in the elderly or focal theta activity over a structural lesion |

Amplifiers are used to magnify the signal for its display. Electrodes are an integral part of the circuit, which consists of the patient and the recording equipment.

Electrode gel provides a contact between the skin and the electrodes and is considered an extension of the electrode. High electrode impedance and loose electrodes can result in electrode "pops," which are spike-like potentials that occur randomly.

Epileptiform activity consists mainly of spikes and sharp waves. Spikes have a duration of less than 70 ms, whereas the sharp waves have a duration of 70–200 ms.

TABLE I.2 Epileptiform rhythms

| Description | Normal | Abnormal |
|-------------|--------|----------|
| *Spikes*: up to 70 ms | Vertex waves, positive occipital sharp transients of sleep | Focal and generalized epileptiform activity |
| *Sharp waves*: 70–200 ms | Benign epileptiform transients of sleep 6/s phantom spike and wave 14-and-6 positive spikes | |

TABLE 1.3  10–20 electrode placement system

| F | Frontal |
|---|---------|
| Fp | Frontopolar |
| C | Central |
| T | Temporal |
| P | Parietal |
| O | Occipital |
| A | Auricular |

Sixteen channels are typically used for a routine scalp EEG. The seventeenth channel is used for electrocardiography (ECG).

Electrodes are placed according to the International 10- to 20-electrode placement system.

The following letters stand for the following electrodes:

As shown in Fig. 1.1, the odd numbers are on the left side of the head, whereas the even numbers are on the right side. The lower numbers are on the anterior side of the head and the higher numbers are on the posterior side. The midline electrodes are indicated by "z" instead of a number.

The electrodes are arranged in montages (arrays), of which there are several, the two most common being the referential (Fig. 1.2) and the bipolar (Figs. 1.3 and 1.4) montages. Changing the montage in an EEG recording will affect the waveforms present due to the change in electrode pairing.

A common referential montage indicates that the reference for each electrode is common; for example, each electrode may be referred to the ipsilateral ear. A bipolar montage means that the reference for one channel is active for the next channel, e.g., F7-T3 means that F7 is the active electrode and T3 is the reference, whereas T3-T5 means T3 is the active electrode and T5 is the reference electrode.

EEG negativity can represent a superficial excitation or a deep cortical inhibition. The upward deflection of a recording is negative, whereas the downward deflection is positive. Spikes originate in the electrode common to a phase reversal. When EEG data was acquired in the analogue mode, a routine EEG recording typically used referential, longitudinal

FIGURE 1.1  Electrode positions

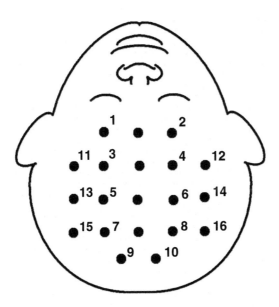

FIGURE 1.2  Referential montage

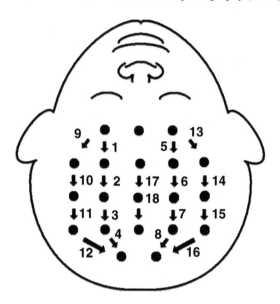

FIGURE 1.3  Longitudinal bipolar montage

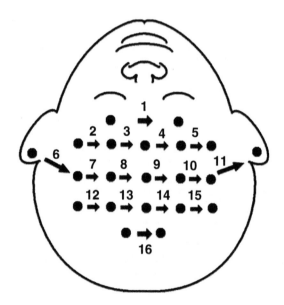

FIGURE 1.4  Transverse bipolar montage

bipolar, and transverse bipolar montages. However, now that EEG data is acquired digitally, data can be reformatted in any chosen montage and viewed in various ways to clarify the appearance of suspicious patterns.

The routine sensitivity used in many EEG laboratories is initially 7 µV. However, the sensitivity can be changed depending upon the amplitude of the EEG recording. The sensitivity can be increased to 5 µV, or 3 µV, or further, if the EEG amplitude is low, as in the elderly; and it can be decreased to 10 µV, 15 µV, or further, if the amplitude is high, as in the young child. Filtering cleans up the EEG tracing, but may remove spikes (at higher filtering) or slow waves (at lower filtering) of pathological interest.

A routine EEG is typically 20 min of artifact-free recording. Filter settings are 1 and 70 Hz.

Hyperventilation, photic stimulation, and sleep are used as activation methods, as these methods promote epileptiform activity in different epilepsy syndromes. Responses are considered abnormal only when there is marked asymmetry or the emergence of epileptiform potentials. Physiological slowing should not outlast hyperventilation by more than 1 min. A 50 % lateralized asymmetry in voltage, or the persistence of slowing, is regarded as abnormal. After a recording made in the wakeful relaxed state, photic stimulation is performed. After photic stimulation, hyperventilation is done for 3 min. In the case of absence seizures, hyperventilation is done for 3 or 5 min. Hyperventilation is contraindicated in patients with recent stroke, carotid artery disease, and severe atherosclerosis. If the routine recording is normal but the suspicion for epilepsy is high, then sleep-deprived EEG may be helpful.

.

# Chapter 2
## Interpretation of EEG

For the interpretation of EEG, the patient's condition is described, including whether the patient is awake, has eyes closed or open, and is relaxed or tense. Activation methods are also described.

A routine EEG recording should include the name of the patient, other identification, date of the study, a clinical diagnosis, names of medications, and date of last seizure.

The following is a sample of a routine recording:

*This recording begins with the patient awake with eyes closed. The posterior dominant rhythm is alpha at 9 Hz, reactive to eye opening. Faster frequencies from the frontal lobes were noted, with a normal anterior-posterior gradient. Photic stimulation and hyperventilation (HV) were performed and did not induce any abnormal activity. Sleep was achieved, with appropriate sleep-related changes. No epileptiform activity was seen.*

*IMPRESSION: This is a normal awake and sleep EEG.*

If the EEG is abnormal, the body of the report should describe the specific abnormalities and the summary should give the electrodiagnostic impression.

The EEG is interpreted in the context of the patient's state. Slow wave activity would be normal in stages III and IV of sleep but is considered abnormal when the patient is awake. While interpreting an EEG study a physician should keep in mind the patient's clinical state, examine the

A.Q. Rana et al., *Neurophysiology in Clinical Practice,*
In Clinical Practice, DOI 10.1007/978-3-319-39342-1_2,
© Springer International Publishing Switzerland 2017

composition of frequencies, and compare the right-to-left symmetry. Background changes in the awake state should be noted. Abnormal responses to photic stimulation and HV should be noted, as should any abnormal wave, such as sharp waves and spikes.

The alpha rhythm is seen in a normal awake individual with eyes closed and in a relaxed state; it is most dominant in the occipital electrodes O1 and O2. The posterior dominant alpha rhythm is usually suppressed by eye opening but returns with eye closure, a response commonly evaluated in routine EEGs.

The amplitude of the posterior dominant alpha rhythm is usually between 15 and 50 μV; however, it may be slightly lower in older individuals. The lower amplitude is not considered abnormal if the background is normal. Asymmetry of background alpha amplitude between hemispheres should be less than 50 %. The bilateral slowing of the posterior dominant rhythm to less than 8 Hz is usually due to diffuse encephalopathy, although the underlying cause cannot be determined solely through EEG.

Interpretation of EEG recordings requires basic knowledge of the possible waveform morphologies and frequencies and the ability to develop a spatial analysis of what is being recorded. These items are discussed below.

The frequency of delta waves is less than 4 Hz, and this is considered normal in sleep. Focal subcortical lesions may demonstrate polymorphic delta activity.

Theta activity of 4–7.99 Hz is normal in drowsiness and as seen in the posterior slow waves of youth, but is abnormal when seen as temporal theta in elderly individuals with focal theta activity over a structural lesion.

Alpha activity of 8–12.99 Hz is normal as the posterior dominant rhythm. Generalized alpha activity is indicative of a poor prognosis, described as an alpha coma. This form of coma is described as anterior alpha dominance without any variation in frequency or amplitude and without reactivity.

Beta activity is more than 13 Hz and this is normal in the anterior region; however, it could be drug-induced, as when it

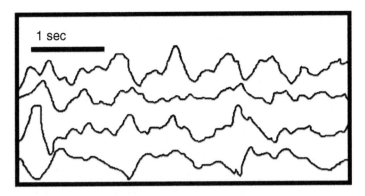

FIGURE 2.1 Diffuse slowing

is due to benzodiazepine or barbiturate use, or it could be a breach rhythm if there is a skull defect (the skull and scalp act as a high-frequency filter). Asymmetry in beta activity with absence on one side may be indicative of a structural cortical or extra-axial lesion on that side, such as cerebral edema or a subdural hematoma.

Slow waves, if focal within a normal background, may be due to structural lesions. Generalized slowing of the background is commonly due to an encephalopathy (Fig. 2.1). Focal slow activity may be recorded over areas of cerebral damage. Focal slowing is least evident when the patient is wakeful and is most apparent in stage I sleep.

Slowing of the background to less than 8 Hz is usually abnormal over the age of 1 year. The alpha rhythm has the highest amplitude in the O1 and O2 electrodes. Theta and delta frequencies are not usually prominent in normal awake EEG. The occipital dominant alpha rhythm attenuates on opening of the eyes or if the patient is tense. Asymmetry of the posterior dominant rhythm should not be more than 50 %; however, less than 50 % asymmetry may be normal. Decreases in the amplitude of alpha activity and slowing of the background may be seen with increasing age in adults.

A normal posterior dominant rhythm is established by the age of 1 year and rapidly gains in frequency during childhood, typically being 10 Hz in adulthood. However, up to 5 % of the adult population has minimal or no clearly identifiable posterior rhythm with alpha frequency range. The slow waves of youth are superimposed on a normal posterior dominant rhythm and are in the delta range, which may be seen up to 30 years of age. Vertex waves are broadly distributed; negative sharp wave transients are most prominent at the Cz electrode and mark the onset of stage I sleep, when they are seen most prominently. Sleep spindles, seen by age 2 years, are 11- to 14-Hz waves and typically of 1–2 s in duration, occurring bilaterally and maximally within the frontocentral region. They are most prominent at the C3 and C4 electrodes and are usually seen in stage II sleep. K-complexes are diphasic vertex waves that appear in association with sleep spindles; the K-complexes may occur before, during, or after the vertex sharp wave. Both sleep spindles and K-complexes are thought to originate within the thalamus.

Drowsiness is characterized by the fallout of the posterior alpha rhythm and the appearance of slower frequencies in the theta bands, seen predominantly in the frontocentral regions (Fig. 2.2). Occasionally, alpha activity may persist in stage I sleep. Positive occipital sharp transients (POSTs) are "sail-shaped" morphologies that are surface-positive; they occur bisynchronously over the occipital regions during drowsiness and early stages of sleep. POSTs are most often seen in young adults.

In stage II sleep, sleep spindles and K-complexes appear (Figs. 2.3 and 2.4). Deeper stages of sleep may not be seen in most routine EEGs. In stage II sleep, theta activity is further increased, with sleep spindles in the frontocentral regions. Delta activity may appear in 20 % of the sleep recordings. In stage III sleep, delta activity increases to almost half of the EEG recording, with spindles persisting. Stage IV sleep is defined as delta activity greater than 50 %, with the fading of sleep spindles.

TABLE 2.1 Normal sleep EEG in stages I and II

| Name | Characteristics |
|------|-----------------|
| Sleep spindles | 1- to 2-s duration, frequency of 11 to 14 Hz, maximally seen at C3 and C4 electrodes, seen in stage II sleep |
| Vertex waves | Maximally seen at Cz electrode, during stage I and II sleep |
| K-complexes | Seen in stage II sleep |
| Positive occipital sharp transients of sleep (POSTs) | Maximally seen at O1 and O2 electrodes in stage II sleep |

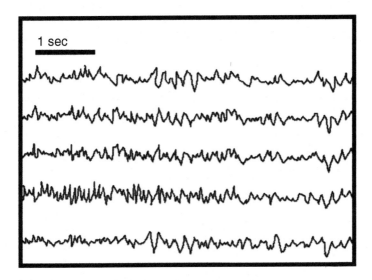

FIGURE 2.2 Frontal theta activity (drowsiness)

REM (rapid eye movement) sleep is indicated by low-voltage fast activity. A 6- to 8-Hz rhythmic activity may appear in the vertex and frontal regions. Lateral eye movements result in an artifact that is seen in the anterior leads of all EEG recordings.

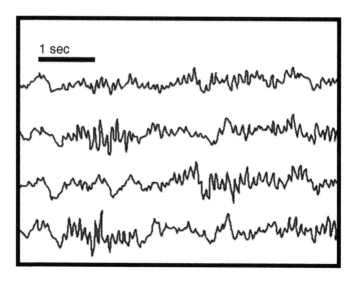

FIGURE 2.3  Sleep spindles

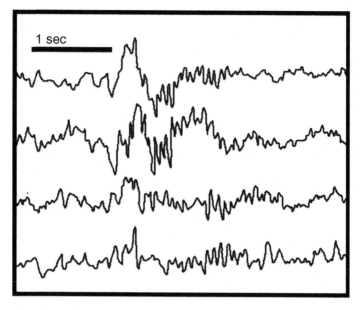

FIGURE 2.4  K-complexes

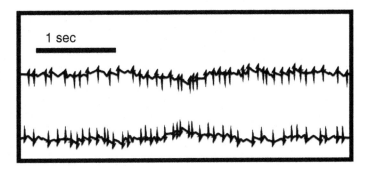

FIGURE 2.5  Muscle artifact

Electromyography (EMG) activity may be seen and is easily differentiated from EEG activity (Fig. 2.5). EMG activity is most prominent in the awake state and is characterized by short-duration spikes with amplitudes of 50 μV, most prominent in the frontal and temporal regions. EMG discharges are not accompanied by slow waves and may be attenuated if the patient relaxes and opens their mouth.

Occasionally a *Mu* rhythm, classified as a central rhythm in the alpha frequency, may be seen in normal EEG recordings. It has a frequency in the range of 8–10 Hz. It is negative and of rounded morphology. *Mu* rhythms are most prominent in the central regions around the C3 and C4 electrodes and are attenuated by movement, or contemplation of movement, of the contralateral limb.

Other electrographic features that are important to differentiate from epileptiform activity include the following. *Lambda waves* are positive waves seen in the occipital region in the awake state and they disappear with eye closing. *Wicket spikes* are sharp waves in the temporal region seen during drowsiness and light sleep. They resemble *Mu* patterns, but have a more striking high-voltage spike-like morphology. Wicket spikes can be differentiated from epileptic spikes by the absence of an after-going slow wave. *Fourteen- and 6-Hz positive spike activities* are sharply contoured waveforms seen in the temporal region. These occur mainly in drowsiness and light sleep. *Benign epileptiform transients of sleep* (BETS) are small spike-like potentials seen in drowsiness and light sleep. They are characterized as being less than 50 μV in amplitude and less than 50 ms in duration and are prominent in the temporal regions. They are differentiated from epileptic spikes by their short duration, normal EEG background, small amplitude, and no after-going slow-wave. BETS are usually seen in children and have no correlation with epilepsy. Frontal intermittent rhythmic delta activity (FIRDA; Fig. 2.6) should prompt investigations for toxic/metabolic encephalopathy, but does not suggest an epileptic predilection.

TABLE 2.2 Normal EEG variants

| Name | Characteristics |
| --- | --- |
| *Mu* rhythm | Wicket-shaped spikes of <1 s duration, frequency of 10 Hz, most prominent at C3 and C4 electrodes and blocked when the individual is thinking about the motion of a limb |
| Lambda waves | Positive occipital waves, disappear on closing of the eyes |
| Wicket spikes | Temporal sharp waves and spikes, seen during light sleep and drowsiness |
| 14- and 6-Hz positive spikes | Sharply contoured spikes with temporal predominance; occur during light sleep and drowsiness |
| Benign epileptiform transients of sleep (BETS) | Small spike-like potentials in the temporal region, seen during drowsiness and light sleep |
| Rhythmic temporal theta of drowsiness | Rhythmic prolonged burst of sharply contoured theta waves with temporal predominance, seen in drowsiness |

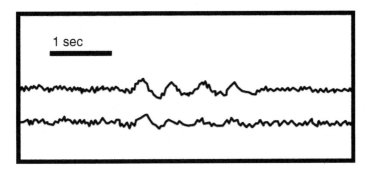

FIGURE 2.6  Frontal intermittent rhythmic delta activity (FIRDA)

Activation methods commonly used during routine EEG recordings include HV and photic stimulation. Hyperventilation typically results in bilaterally synchronous slow waves, often with sharp contours. This activity is maximal posteriorly in the posterior head regions in children, but is maximal in the frontal regions in teenagers and adults. The normal response ends within 1 min of the cessation of HV. With HV, there is a generalized slowing of the background to the theta range in both hemispheres. However, absence of slowing is not abnormal. The electroencephalographic finding characteristically seen after hyperventilation in about 50 % of children with cerebrovascular disease includes gradual frequency decrease and activation of amplitude of slow waves which appear after the disappearance or attenuation of ordinary build up. This is termed the "re-build up" phenomenon, which is rarely seen and therefore may be under-recognized.

An asymmetry of the HV response is abnormal and suggests an abnormality on the side with higher amplitudes. The HV technique serves to augment the 3-Hz spike-and-wave discharges seen in typical childhood absence epilepsy. Similarly, HV can activate the slow spike-and-wave pattern of Lennox-Gastaut syndrome (LGS) in 50 % of patients (Fig. 2.7). HV can also augment focal slowing due to subcortical dysfunction.

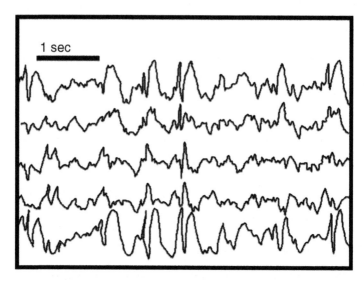

FIGURE 2.7 Spike-and-wave complex (Lennox-Gastaut syndrome)

Photic stimulation may activate epileptiform discharges in some forms of primary generalized epilepsy, such as juvenile myoclonic epilepsy (JME). A train of light flashes is run in front of the patient with closed eyes for 10 s; the trains are delivered every 20 s , with an initial rate of 3 per s, slowly increasing to 5, then 7, and up to 30 per s in some laboratories.

Visual evoked responses, driving responses, and the photo-myoclonic response (PMR) may be seen as normal conditions with photic stimulation. The driving response is time-locked to the stimulus, whereas the visual evoked response occurs approximately 100 ms after the stimulus. The driving response is also present at faster frequencies. The time lapse between the appearance of the visual evoked response and the response seen only at lower frequencies differentiates the visual evoked response from the driving response to photic stimulation.

The PMR is characterized by contractions of the frontalis muscles, time-locked with a delay of 50–60 ms, and is suppressed by eye opening. The PMR consists of brief repetitive muscle spikes; it is thus not cortically generated and is anterior in location, whereas photoconvulsive responses (PCRs) are generalized or posterior.

The PMR must be distinguished from the PCR in that the PMR stops at the end of the stimulus, whereas the PCR outlasts the stimulus.

The PMR has the same frequency as the flash frequency, whereas the PCR is slower, usually in the range of 3 per s. The photoconvulsive discharges are spike-and-wave cortical discharges. The PCR is more likely to be epileptic-associated if this response continues after the end of the flash.

# Chapter 3
# EEG Patterns in Seizure Disorders

Patients having EEG studies for a seizure disorder should be studied when awake as well as while asleep in the same record, since interictal epileptiform activity is increased during sleep and when drowsy . Patients are routinely asked to arrive for the examination after having been sleep-deprived the night before to allow for easy transition to drowsiness.

If the routine recordings are normal and suspicion for epilepsy is high, then prolonged monitoring with sleep deprivation should be done. Patients are usually kept awake for 24 h before the EEG. The transition from drowsiness to light sleep may be very short, and therefore must be examined closely for the detection of abnormal epileptiform activity. Sometimes sharply contoured slow waves may be misinterpreted as spikes or sharp wave activity, particularly in young children. False-positive EEG interpretations are to be avoided. Most EEG phenomena noticed by beginners are artifacts.

True epileptiform spikes are usually stereotypical and stand out from the background with a fast rising phase. They may be followed by a slow wave and have a potential field of activity. A *sharp transient* is any wave of any duration that has a pointed peak at standard recording speeds. As such, any sharply contoured waveforms that do not meet the criteria of being epileptiform are called sharp transients. These sharp transients are often variable in morphology; their rising phase may be slower than the falling phase, and the rising phase is

A.Q. Rana et al., *Neurophysiology in Clinical Practice*,
In Clinical Practice, DOI 10.1007/978-3-319-39342-1_3,
© Springer International Publishing Switzerland 2017

usually not followed by slow waves. Often they do not have a potential field; there is no change with sleep see Table 3.1 below for more details.

TABLE 3.1 Differences between spike and nonspike potentials

| Spike potential (epileptiform) | Non-spike potential (sharp transient) |
| --- | --- |
| Stereotypical in appearance | Variable in morphology |
| Usually present consistently | Nonconsistent presence |
| Rising phase is fast | Rising phase is slower than the falling phase |
| Usually followed by a slow wave | Usually not followed by a slow wave |
| Stands out from background | Does not stand out from background |
| Activated in sleep state | No change with sleep state |
| Defined potential field | Does not have a defined field, may be a single electrode |

Spikes and sharp waves are abnormal in most conditions, except when they are in the form of vertex sharp waves and positive occipital sharp transients of sleep (POSTS) in stage I sleep, 14- and 6-Hz positive spikes, wicket spikes, occipital lambda waves, or six-per-second phantom spike and wave discharges. Spikes are between 20 and 70 ms in duration and sharp waves are 70–200 ms in duration. Spike and wave discharges correlate more with the likelihood of epilepsy than a single spike. *Generalized* spike and wave discharges are of the following types see Table 3.2:

1. Three-per-second spike and wave
2. Slow spike and wave
3. Fast spike and wave

   *Absence epilepsy* (Fig. 3.1) has classic three-per-second spike and wave complexes, which increase with hyperventilation with a normal background. Absence epilepsy is an example of a primary generalized epilepsy.
   *Juvenile myoclonic epilepsy (JME)* is another example of a primary (i.e., idiopathic) generalized epilepsy syndrome, in which the background EEG is normal, but in

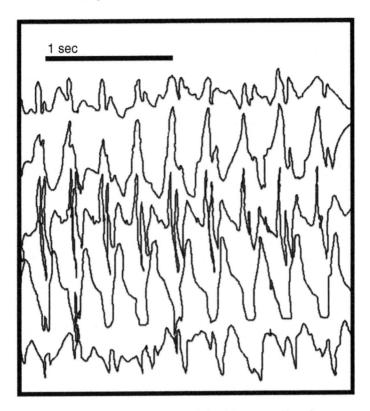

FIGURE 3.1   3-Hz spike and wave activity (absence epilepsy)

which faster (>3.5 Hz) generalized polyspike discharges with slow waves occur, more prominent in the fronto-central regions. The condition is characterized as myoclonic jerks in the early morning, with the possibility of generalized tonic-clonic seizures. This condition typically requires lifelong therapy.

*Generalized tonic or tonic/clonic seizures* (GTCS; Fig. 3.2) are characterized by a progressively higher amplitude and lower frequency discharge pattern observed simultaneously in both cortical hemispheres, reaching a maximum of 10 Hz.

*Partial/focal epilepsies, those related to isolated cerebral dysfunctions* are associated with variable focal EEG

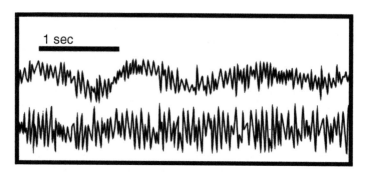

FIGURE 3.2 Generalized tonic seizure

patterns. They may be unilaterally temporal or frontal in location with focal slowing depending on the site of epileptic origin. Epileptiform discharges, either sharp waves or spike waves, are typically seen in these areas of focal slowing and tend to be most evident in stages III and IV of non-rapid eye movement (REM) sleep. The EEG can often be normal, particularly if sleep is not obtained. Focal spikes indicate a focal or partial epilepsy syndrome. Focal spikes are considered important if they are consistent within a definable field and there is no artifact. Single spikes during routine EEGs may not be significant see Table 3.3.

TABLE 3.2 Generalized spike discharges

| Pattern | Features | Significance |
|---|---|---|
| Burst suppression | Bursts of slow waves with superimposed sharp waves, mixed with periods of relative flattening | Severe anoxic or diffuse encephalopathy with corresponding poor prognosis |
| Three-per-second spike wave | Spike or polyspike complexes at a rate of 2.5–4.0/s and increased by hyperventilation | Generalized epilepsy 3-Hz spike and wave indicative of absence epilepsy |

TABLE 3.2 (continued)

| Pattern | Features | Significance |
|---------|----------|-------------|
| Six-per-second (phantom) spike wave | Small spike-wave complexes at brief bursts of alpha-theta usually lasting less than 1 s. Usually appear in adolescents and young adults in drowsiness. Disappear in sleep. Frontal or occipital predomi nance | Frontal wave is associated with generalized tonic-clonic seizures, whereas occipital wave may not be associated with seizures |
| Slow spike wave | Spike wave at a rate of 1.0–2.0/s | Lennox-Gastaut syndrome or generalized seizures |
| Hypsarrhythmia | High-voltage bursts of theta and delta with multifocal sharp waves mixed with periods of relative suppression | Infantile spasms |

Partial epilepsies may also be idiopathic or symptomatic, depending on the absence or presence of an anatomic abnormality. Examples of idiopathic localization-related epilepsies include benign rolandic epilepsy (BRE) and benign occipital epilepsy (BOE). Rolandic epilepsy is characterized by stereotypical centrotemporal spikes and triphasic morphology, maximal at electrodes C3 and C4, followed by a slow wave. The findings are activated by sleep. Benign occipital epilepsy (BOE), a similar benign age-related localization-related epilepsy syndrome, is characterized by rhythmic occipital spikes, often moderately lateralized, which "block" with eye opening. These EEG patterns are localized as described, but during established sleep can be become more generalized in distribution. Background EEG in both conditions is otherwise normal.

TABLE 3.3  Focal spike discharges

| Type | Features | Significance |
| --- | --- | --- |
| Occipital spikes | Mainly negative or biphasic spikes over the occipital region | May indicate benign occipital epilepsy, although occipital spikes may not always be benign |
| Rolandic spikes | Spike and slow wave complex, triphasic in appearance, with maximum presence at C3 and C4 | Indicate benign rolandic epilepsy, although may be seen in individuals without seizures |
| Temporal sharp waves | May be seen in the anterior, mid, or posterior temporal regions | Associated with partial complex seizures, generalized tonic-clonic seizures, or psychological complaints |
| Periodic lateralized epileptiform discharges (PLEDs) | Unilateral or bilateral sharp and slow wave complexes at a rate of 1–2/s | Anoxia, herpes simplex encephalitis, stroke, tumor, or any other destructive process |

Triphasic waves (Fig. 3.3) are high-amplitude ($>70$ µV) positive sharp transients that are preceded by negative waves of low amplitude with bifrontal predominance at 1–2 Hz; they are seen classically in hepatic encephalopathy, but are also associated with multiple toxic/metabolic and even structural lesions.

Epileptiform discharges may occur in a regular, i.e., periodic fashion. Periodic epileptiform discharges are sharp waves, which are of high amplitude and occur at a rate of 0.5–3/s (Fig. 3.4).

Periodic discharges are seen when there is destruction of brain parenchyma, such as with stroke, abscess, herpes encephalitis, or brain tumors. These are termed secondary epilepsies, since there is an underlying cause. They have amplitudes of 100–300 µV with an early negative component followed by a positive phase. A term often used by electroencephalographers to describe this appearance is PLEDs (periodic lateralized epileptiform discharges) as seen in the conditions

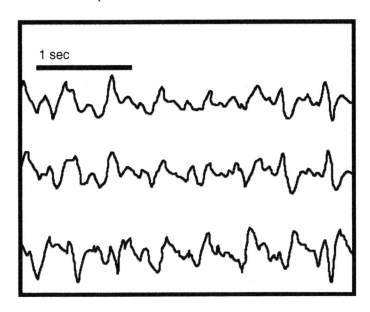

FIGURE 3.3  Triphasic waves

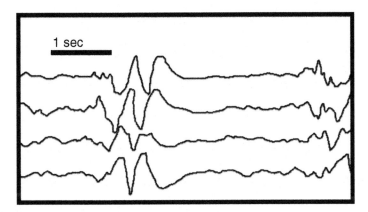

FIGURE 3.4  Burst of sharp and slow wave complexes (seen in subacute sclerosing pancreatitis [SSPE])

mentioned above (Fig. 3.5). PLEDs are to be distinguished from generalized periodic epileptiform discharges (GPEDs ), as seen in more diffuse disorders, such as Creutzfeldt-Jakob disease (CJD) see Table 3.4 for more details.

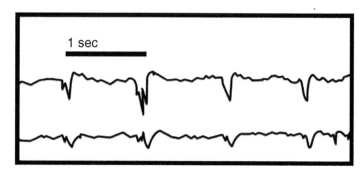

FIGURE 3.5  Periodic lateralized epileptiform discharges (PLEDs)

TABLE 3.4  EEG patterns of disorders with frequent spikes

| Type of disorder | Characteristics |
| --- | --- |
| Rolandic epilepsy | Stereotypic spikes and slow wave with normal background, triphasic in appearance, maximal presence at C3 and C4 and activated by sleep |
| Occipital epilepsy | Occipital spikes, increased by light sleep and suppressed by eye opening |
| Absence epilepsy | Classic three-per-second spike-wave complex, increased by hyperventilation |
| Juvenile myoclonic epilepsy | Generalized polyspike and slow wave discharges with normal background, and higher amplitude in the frontal region |
| Complex partial seizures | Unilateral temporal or frontal spikes depending upon the site of origin of epileptiform activity |
| Simple partial seizures | Midline spikes that may have a negative phase or may be biphasic |
| Generalized tonic-clonic seizures | Generalized polyspike and slow wave pattern, although ictal activity may not have slow waves |
| Creutzfeldt-Jakob disease | Periodic complexes with a duration of 0.5–2.0/s and maximal anteriorly, background slowing with low voltage, may fade away during sleep |

TABLE 3.4 (continued)

| Type of disorder | Characteristics |
|---|---|
| Herpes simplex encephalitis | Generalized slowing with irregular periodic complexes of sharp waves or slow waves |
| Anoxic encephalopathy | Variable from normal to isoelectric depending upon severity of underlying condition. Burst suppression may be seen. Alpha coma indicates a poor prognosis |
| Lennox-Gastaut syndrome | Slow spike-wave complex, with slowing and disorganization of background |
| Subacute sclerosing panencephalitis | Periodic high-amplitude bursts of sharp waves of 0.5–2.0 s duration, superimposed with an irregular delta wave in both hemispheres. Seen as a long-term sequela of prior measles infection |

Lennox Gastaut syndrome is a catastrophic childhood epilepsy syndrome characterized by multiple seizure types, mental retardation, and behavior problems. There is often poor seizure control. The EEG shows generalized slow (<2.5 Hz) spike-wave complexes against a slow background. This condition is therefore an example of a symptomatic generalized epilepsy syndrome.

Patients with hypoxic ischemic encephalopathy have diffuse slowing of background, as well as background disorganization. A burst suppression pattern can be seen in patients with severe anoxic encephalopathy, or as a consequence of induced anesthesia.

In cases of possible brain death, special techniques are used to ensure that any existing cerebral electrical activity is detected. Specific requirements include the number of electrodes used, electrode impedance, amplitude requirements, patient core body temperature, and lack of any comorbidities that may hinder cortical activity.

In summary, EEG is a remarkably useful diagnostic test when appropriately performed and utilized in the correct clinical contexts. A single routine EEG recording will confirm the diagnosis of epilepsy in 60 % of persons with this condition. The specific EEG pattern can help to accurately identify the specific epilepsy syndrome, thereby permitting more accurate prognostication and treatment. An EEG recording can identify the presence of nonconvulsive status epilepticus in patients with altered mental status when there are no otherwise obvious clinical features. An EEG recording can also be useful in evaluating states of altered mental status such as acute encephalopathies or more chronic conditions such as neurodegenerative conditions, including CJD. More sophisticated techniques, such as video-EEG recording, can allow clinical correlations and can facilitate the identification of psychogenic nonepileptic seizures. Intracranial EEG recordings are often performed when accurate localization of seizure foci is required and this is not possible with surface recordings, particularly when epilepsy surgery is being contemplated.

In conclusion, despite EEG being a century old, it remains an invaluable and in fact indispensable technique in the diagnostic armamentarium of the twenty-first century physician.

# Part II
# Electromyography and Nerve Conduction Studies

There is much to know about the rapidly expanding field of electrodiagnostic medicine, and there are several excellent texts and periodicals that provide up-to-date information that are highly valuable. However, when the amount of information is extensive, it is challenging to the beginners and the novice to navigate.

The purpose of this manual is to meet the day-to-day needs of the health care professional, particularly the trainees, to acquire the core knowledge and skills that are the essential components of a reliable electrodiagnostic evaluation. This book is specifically designed for trainees in neurology and in physiatry to have a readily available resource in their pocket to be able to deal with common and uncommon challenges in the evaluation and diagnosis of neuromuscular disorders.

Obtaining an appropriate history and undertaking a relevant clinical examination are important aspects of any patient's assessment. The clinical evaluation is essential in generating a differential diagnosis and guiding the electromyographer to undertake appropriate electrophysiological tests. Electrodiagnosis, hence, is considered an extension of the physical examination. Electrophysiological evaluation then becomes an important tool in the evaluation, diagnosis, prognosis, and management of neuromuscular problems.

Electrodiagnostic studies for the purpose of this book entail the evaluation of the lower motor neuron including the muscle. Electrodiagnostic studies can evaluate the entire motor unit that includes the anterior horn cell, roots, axon, myelin, neuromuscular junction, and the muscle.

An adequate electrodiagnostic evaluation will require knowledge of the anatomy of nerves and muscles, electrophysiology, pathophysiology of the disease conditions being studied, the clinical presentations, and the technical issues.

In summary, the essential components of an electrodiagnostic evaluation, hence, are the following:

1. Evaluating the patient by obtaining an adequate history and undertaking a physical examination with a view to developing a differential diagnosis.
2. Selecting an appropriate set of electrodiagnostic tests and explaining them to the patient.
3. Performing electrodiagnostic tests in a competent fashion, understanding the technical limitations and the patient's ability to tolerate the test.
4. Interpreting the results in view of the electrophysiological information, technical limitations, history, and physical examination to narrow the differential diagnosis, arrive at a specific diagnosis, or rule out certain important medical conditions and communicating this interpretation to the referring physician.

# Chapter 4
## Clinical Evaluation

Electrophysiological techniques, including nerve conduction and electromyography, are useful in the evaluation of weakness, paralysis, muscle atrophy, pain, numbness, tingling, burning, fatigue, stiffness, cramps, gait imbalance, and other conditions. Hence, these techniques are an extension of the clinical examination, where a nerve, muscle, or any component of the motor unit needs to be evaluated. The electrophysiological test has to answer a clinical question.

As in any clinical situation, relevant history has to be obtained, along with a detailed physical examination, to be able to develop a differential diagnosis. Appropriate laboratory tests are then arranged to determine the likelihood of a disease or condition being present or absent.

Electrodiagnostic studies are undertaken for the purpose of diagnosis, to establish the severity of the diagnosed disease condition, to understand the progression of the condition, to define prognosis, to determine the localization of the lesion, and to initiate an appropriate treatment protocol.

A.Q. Rana et al., *Neurophysiology in Clinical Practice*,
In Clinical Practice, DOI 10.1007/978-3-319-39342-1_4,
© Springer International Publishing Switzerland 2017

# Precautions and Pitfalls

## *Patient Considerations*

- An electrophysiological test should proceed with an appropriate explanation and appropriate consent from the patient.
- The patient has to be comfortably gowned.
- The room should be comfortable and the privacy of the patient respected.
- Reassure the patient that the test will stop at any time when requested.
- A patient who has received an explanation of the procedure is more likely to comply.

## *Adequate History and Examination*

- Generate the differential diagnosis to be able to select the most appropriate set of tests.
- History of infection transmission (human immunodeficiency virus, hepatitis)
- Medication history for:
    - Drugs that can cause bleeding (e.g., warfarin, aspirin, nonsteroidal anti-inflammatory drugs)
    - Drugs that interfere with electrophysiology (e.g., pyridostigmine)

# Cautions and Complications (see Table 4.1)

## *Specific Patient Conditions*

Lymphedema: Risk of cellulitis and chronic serous drainage.

Prosthetic joints: Risk of prosthetic joint infection declines rapidly in the first few months and continues to decline.

Electrically sensitive patients: Patients with cardiac catheters, intravenous and intraarterial lines, and those with

TABLE 4.1

| Arising from needle electromyography | Bleeding, such as with the use of anticoagulants |
|---|---|
| | Infection more common among immuno-compromised care to maintain sterile needle |
| | Penetration of tissue (nerve, lung, blood vessel) |
| | Disease transmission (hepatitis, human immunodeficiency virus) |
| | Cellulitis, open or infected skin site |
| Arising from electric stimulation | Ventricular fibrillation |
| | Automatic implanted defibrillators |
| | Cardiac pacemaker (not over or across site) |
| Infection control | Use of protective barriers |
| | Sterile electrodes |
| | Disposable electrodes |
| | Aseptic cleaning techniques |
| Disposal | Regulated waste disposal and proper housekeeping |

pacemakers and pacing leads. Any electrical stimulation should be performed in a manner that avoids the possibility of current transmission.

"Leakage current" to the instrument chassis and then delivered to an improperly grounded patient should be avoided.

Bleeding disorders: Patients with known disorders of hemostasis or those on major anticoagulants.

Pregnancy: No reported complication.

# Further Reading

Nora LM. American Association of Electrodiagnostic Medicine Guidelines in Electrodiagnostic Medicine: implanted cardioverters and defibrillators. Muscle Nerve. 1996;19:1359–60.

Referral guidelines for the electrodiagnostic medicine consultation. Approved by the American Association of Neuromuscular and Electrodiagnostic Medicine (formerly AAEM): August 1996.

# Chapter 5
# Essential Neurophysiology

Neurophysiology is physiology and neuroscience that focuses on the analysis of nervous system function, and electrophysiology is the study of the electrical properties of biological cells and tissues. Electrophysiology involves measurements of voltage change or electric current on different scales, from single ion channel proteins to large tissues such as muscle. Recordings of electric signals from the nervous system may also be referred to as electrophysiological recordings.

## The Motor Unit

Most mature extrafusal skeletal muscle fibers are innervated by a single α motor neuron. Because of the presence of more muscle fibers than motor neurons, single motor axons branch within muscles in order to synapse on a variety of diverse fibers that are commonly dispersed over a relatively wide area within the muscle. This specific layout decreases the probability that damage to a single or a few α motor neurons will considerably modify a muscle's action.

Since an action potential produced by a motor neuron typically brings all of the muscle fibers in its association to a threshold, a single α motor neuron and its accompanying muscle fibers collectively construct the lowest possible force that can be elicited to generate movement.

A.Q. Rana et al., *Neurophysiology in Clinical Practice*,
In Clinical Practice, DOI 10.1007/978-3-319-39342-1_5,
© Springer International Publishing Switzerland 2017

Hence, each motor nerve fiber and the group of muscle fibers that it supplies is a functional entity, since each time the nerve fiber discharges an impulse, the muscle fibers of the whole assembly would be collectively excited. Sherrington was the earliest individual to observe this noteworthy correlation between an α motor neuron and the muscle fibers it innervates, consequently coining the term "motor unit."

Both the motor units and the α motor neurons themselves vary in size, and they also differ in the types of muscle fibers that they innervate. There are three classes of motor units:

1. *Small motor units* innervate small muscle fibers that contract slowly and create rather small forces; however, because of their high myoglobin content, large number of mitochondria, and rich blood supply, these tiny "red fibers" are not susceptible to fatigue. These motor units are known as slow (S) motor units; they have a low threshold for activation, are tonically active, and are notably vital for processes that demand continuous muscular contraction, such as sustaining an upright posture. For example, a motor unit within the soleus, a muscle that is crucial for posture, contains tiny slow units and has an average mean innervation ratio of 180 muscle fibers per motor neuron.

2. *Larger α motor neurons* innervate larger muscle fibers that produce more force; these fibers, however, possess scarce numbers of mitochondria, resulting in fatigue that occurs rapidly. These large, fast motor units, are also known as fast fatigable (FF) motor units and reach their threshold only during rapid movements requiring substantial amounts of force. These units are considerably significant for momentary exertions involving large forces, such as running or jumping. An example of FF motor units is their presence in sprinters, whose legs possess greater amounts of powerful but rapidly fatiguing pale fibers compared with the legs of long distance, marathon runners. A great example is also evident in the motor units of the gastrocnemius, a muscle with both small and large units, which has an innervation ratio of 1000–2000 muscle fibers for each motor neuron; this can produce forces necessary for immediate changes in body position.

3. *Fast fatigue-resistant (FR) motor units are the third class of motor units.* FR units have characteristics that lie between those of classes 1 and 2 above. They are intermediate in size and are not as rapid as FF units. But they are also significantly more resistant to fatigue than the slow motor units, as hinted at by their name. Furthermore, they produce almost double the amount of force of a slow motor unit.

Other distinct variations in motor units are correlated with the highly specialized activity of specific muscles. For example, for the eyes, which require fast movement along with precision, but minimal strength, extraocular muscle motor units are remarkably tiny in size (innervation ratio of three muscle fibers per motor neuron!), and possess copious amounts of muscle fibers that can contract with maximal velocity.

# Synapse

The contact region between an axon and a muscle fiber is called a synapse or neuromuscular junction. The synapse has particular structural and functional features that allow the impulse from the nerve fiber to be transmitted to the muscle fiber through a chemical neurotransmitter link, acetylcholine. Each muscle fiber has one synapse and is innervated by only one motor neuron.

# Synaptic Delay

Synaptic delay is the time necessary for the transmission of an impulse across a synapse. This refers to the interval between the time that the end of a presynaptic fiber experiences the arrival of a new impulse and the initiation of the postsynaptic potential. In synapses that have chemical transmission mechanisms, this delay lasts from 0.3 to 0.5 milliseconds to several milliseconds. For most of this period, mediators are released by presynaptic endings that are governed by nerve impulses.

# Motor Unit Action Potential (MUAP) (see Figs. 5.1– 5.3)

The MUAP is the compound action potential of a single motor unit whose muscle fibers lie within the recording range of an electrode. For quantitative analysis, the following features are observed:

(a) Amplitude (peak-to-peak in millivolts [mV] or microvolts [μV])
(b) Duration in milliseconds (ms)
(c) Number of phases, i.e., monophasic, biphasic, triphasic, or polyphasic
(d) Polarity of each phase, i.e., negative or positive
(e) Number of turns are number of times the action potential wave crosses the baseline
(f) Variation of shape if any, with consecutive discharges
(g) Presence of satellite (linked) potentials, if any

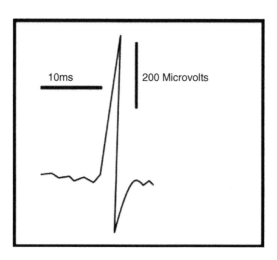

FIGURE 5.1 Normal motor unit potential

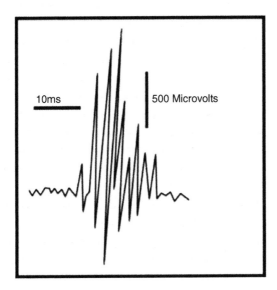

FIGURE 5.2 Neuropathic motor unit

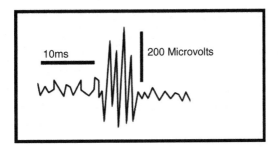

FIGURE 5.3 Myopathic motor unit

## Motor Unit Number Estimate

The motor unit number estimate (MUNE) is a quantitative technique for determining the number of functioning motor units within a muscle. This can be obtained by a variety of methods that include incremental motor nerve stimulation and spike-triggered averaging, or by statistical techniques.

## *Polyphasic Motor Unit Action Potentials*

A MUAP is called polyphasic if it has four or more phases. A phase is defined as the area of an action potential on either side of the baseline. If the potential has multiple reversals of direction without crossing the baseline, it is called a complex or serrated potential. Some polyphasic potentials may have a late satellite or parasite component.

Polyphasic motor unit potentials (MUAP) are seen in reinnervation. The earliest polyphasic potential is low amplitude and short duration, and subsequently, as the muscle fibers hypertrophy, the amplitude is increased in size and so also is the duration. Polyphasic MUAP are also seen in myopathic conditions and characteristically are of low amplitude and short duration.

Polyphasic potentials must be differentiated from doublets, multiplets, tremulousness, and long-duration MUAP.

# Fibrillation Potential

Fibrillation is the spontaneous contraction of individual muscle fibers. Any muscle fiber that is not innervated can be expected to fibrillate. Fibrillation potential (FP) is the action potential of a single muscle fiber occurring spontaneously or after the movement of a needle electrode. These are biphasic or triphasic spikes of short duration, of less than 5 ms, with an initial positive deflection and peak-to-peak amplitude of generally 200 μV to a maximum of 1 mV. These spikes have an associated high-pitched regular sound, described as "rain on a tin roof."

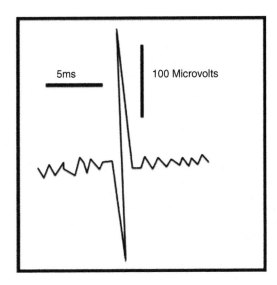

FIGURE 5.4  Fibrillation potential

Positive sharper waves may also be recorded from muscle fibers when the potential arises from an area immediately adjacent to the needle electrode. FPs need to be differentiated from end-plate noise and end-plate spikes and from short-duration and positive-configuration MUAP.

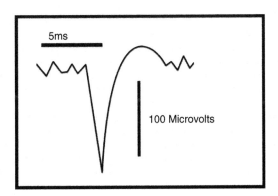

FIGURE 5.5  Positive sharp wave

Positive sharp wave FPs are seen in:

1. Neurogenic conditions with a lesion of the anterior horn cell or lesion of the axon (root, plexus, or peripheral nerve lesion)
2. Neuromuscular transmission diseases, such as myasthenia gravis and botulinum toxin-related conditions
3. Muscle diseases such as myositis, myopathies, and dystrophies, or muscle trauma

## Myotonic Discharges

Myotonic discharges are repetitive discharges that occur at a rate of 20–80 Hz. They may be biphasic (positive-negative) spiked potentials less than 5 ms in duration resembling FPs, or they may be like positive sharp waves of 5–20 ms duration, resembling positive sharp waves. These discharges are recorded after needle electrode insertion is done after voluntary muscle contraction or after muscle percussion. They are a result of independent repetitive discharges of single muscle fibers. The amplitude and frequency of the potentials must both wax and wane, producing a characteristic musical sound frequently likened to the sound of a "dive bomber."

Myotonic discharges are found in the following myopathies:

- Myotonic dystrophy
- Myotonia congenita
- Paramyotonia
- Hyperkalemic periodic paralysis
- Polymyositis
- Acid maltase deficiency

# Fasciculation

Fasciculation is the random spontaneous twitching of a group of muscle fibers within a single motor unit. As a result, these twitches may create movement of the overlying skin or mucous membrane (such as in the tongue). Any electrical activity affiliated with the twitch is known as a fasciculation potential. Furthermore, the electrical activity accompanied by the fasciculation has the configuration of an MUAP. The discharges may arise from any portion of a lower motor neuron, from the cell body to the nerve terminal. These responses may occur as a single fasciculation potential or as a group discharge, referred to as a brief repetitive discharge.

Fasciculation potentials are commonly seen in:

- Normal healthy persons, as benign fasciculations with or without cramps
- Lower motor neuron lesions such as amyotrophic lateral sclerosis (ALS), root lesions, or peripheral neuropathy
- Metabolic conditions such as tetany, thyrotoxicosis, and the effects of anticholinesterase medications

## M-Wave

The M-wave is a compound MUAP evoked from a muscle by electrical stimulation of its motor nerve. This wave is elicited by a supramaximal stimulation to the motor nerve. Common measurements include latency, amplitude, and duration. Normally the configuration is biphasic and stable with repetitive stimulation at low rates, such as 1–5 Hz.

## F-Wave

The F-wave is an action potential that is induced intermittently from a muscle via supramaximal electrical stimulation of the nerve; it is due to antidromic activation of the motor neurons. Its amplitude is approximately 1–5% that of the M-wave and has a variable configuration. The latency is longer than that of the M-wave and is variable.

## H-Wave

The H-wave is a compound MUAP associated with a consistent latency recorded from muscles after stimulation of the nerve. Compared with the M-wave of the same muscle, it has a longer latency. This is most reliably elicited from the gastroc/soleus complex, with an ideal stimulation of long duration (500–1000 ms). When the intensity of the stimulus is adequate, an M-wave of maximal amplitude can be produced, which can reduce the H-wave or suppress it completely. It is speculated that this is a spinal reflex that is associated with electrical stimulation of afferent fibers in a mixed nerve. Primarily through the monosynaptic connection in the spinal cord, there is also motor neuron activation to the muscle. A longer latency is evident with more distal sites of stimulation.

## Recruitment

Motor unit potentials (MUAP) are under voluntary control, and when activated, the firing pattern of the MUAP is assessed in terms of recruitment. Recruitment is the initiation of firing of additional motor units. Normal recruitment occurs initially at low levels of effort, and the firing rates are relatively slow. With increased effort, the frequency of the firing of the motor unit is gradually elevated when the next motor unit is recruited. The motor units are capable of firing within 5–15 Hz for a normal muscle during mild contraction.

Recruitment can also be defined as the ratio of the rate of firing of the individual motor unit to the number of motor units that have been activated. The normal recruitment ratio is less than 5:1. If the ratio is more than 10:1, this indicates a loss of motor units. It is normal for three motor units to be firing at around 15 Hz. However, if there are only two units firing at over 20 Hz, this indicates a loss of motor units as well.

Motor unit potential (MUAP) needs to be evaluated in different areas of the muscle. It is most efficiently evaluated by having the patient maintain a minimal voluntary contraction while the needle electrode is advancing through different areas of the muscle. When the motor units are firing rapidly, and when only a few motor units are present, this indicates the loss of functioning MUAP. The most reliable way of judging the loss of MUAP is by assessing the ratio of the rate of firing of the individual motor unit to the number of motor units that have been activated during mild-to-moderate contraction. Strong contraction needs to be avoided.

## Cramp Potentials

Muscle cramps are an involuntary painful muscle contraction associated with electrical activity. Cramp discharges are most common, but other types of repetitive discharges may also be seen. When a muscle cramps, the individual potentials resemble the MUAP, but the potentials fire rapidly at 40–60 Hz sensation to cessation.

Cramps are common phenomena in normal persons and typically occur as the forcefully in a shortened position. Generally, increasing numbers of potentials firing at similar rates are recruited as a cramp develops, and they may fire irregularly and drop out as the cramp subsides.

Cramps can be felt as benign nocturnal cramps, and also in salt depletion, pregnancy, uremia, myxedema, or chronic neurogenic atrophy.

## Myokymia

Myokymic discharge is a form of involuntary activity where the MUAPs are constantly firing. This discharge may have a relationship with clinical myokymia. The pattern of firing can be very brief and repetitive, initiated by a single MUAP for a short-lived period at a uniform rate (2–60 Hz). This is followed by another short period of up to a few seconds of silence, accompanied by a repetition of the same sequence for a specific potential that occurs at regular intervals. Myokymic discharges are a subclass of group discharges and repetitive discharges. They may be seen in:

- Multiple sclerosis
- Brain stem neoplasms
- Facial palsy
- Polyradiculopathy
- Plexopathy
- Chronic nerve entrapments

Understanding and interpreting the different spontaneous activities together with interpretation of MUAPs is one of the most important components of an electrodiagnostic study and the chapter provides a basic framework for it.

# Chapter 6
# Instrumentation

All electrodiagnostic instruments, analog or digital, should meet the specifications recommended by the American Association of Neuromuscular and Electrodiagnostic Medicine. Digital instruments allow for the use of automatic cursor placements, amplitudes, area measurements, averaging, frequency analysis, interference pattern analysis, and trigger delay lines; they also have the ability to change the display sensitivity even after the waveform is captured. Embedded programs allow the use of sophisticated functions, such as motor unit estimates and report generation. The instruments should be user-friendly.

To obtain accurate information, the electrodiagnostic technician and consultant should be familiar with the sources and magnitudes of the various equipment errors, and the techniques that can be used to minimize the adverse effects. The electrodiagnostic medicine consultant must know how and when to override the computer-generated response and be in control of the technology.

## Electricity and Charge

The basis of electricity is charge and charge flow. Electrical equipment in much of the world is typically powered by a 110-V, 60-cps line current. Within an electromyography

A.Q. Rana et al., *Neurophysiology in Clinical Practice*,
In Clinical Practice, DOI 10.1007/978-3-319-39342-1_6,
© Springer International Publishing Switzerland 2017

(EMG) machine, the line voltage is lowered to a direct current voltage between 5 and 15 V that is used to power the amplifiers, filters, and computer circuits.

Current is measured in amps and is defined as the amount of charge (coulombs) flowing per unit of time in seconds. The usual charge carriers are negative (electrons).

The bases of instrument design are voltage, current, and impedance. An analogy can be drawn between these entities and the characteristics of water flowing down a river:

Voltage (V): Voltage is the potential energy per unit charge (joules/coulombs) (*steepness of the river grade determines the current*).

Current (C): Current is the charge flow per second (*amount of river water flowing per second*).

Impedance (I): Impedance is the ratio of voltage to current (*friction between the water and the river bed*).

## Filters

The most important function of a filter is noise attenuation. Every electric device that conducts an electric charge changes the signal, that is, filters it. An ideal device should eliminate frequencies that constitute noise while allowing the frequencies that correspond to physiological signals. Filters are designed with capacitors and resistors. Capacitors are devices (like valves) that impede current flow by an amount that depends on the frequency content of the current. Resistors supply a constant impedance to current flow.

*High-Pass Filters*  Allow high frequencies, e.g., 20 Hz, to pass through. They are set to stop low frequencies, done by placing a capacitor in the signal path.

*Low-Pass Filters*  Allow low frequencies to pass through. They are set to stop high frequencies, done by placing a capacitor between the signal path and ground.

# Amplifiers

Amplifiers convert a low-voltage waveform to a higher-voltage waveform. Two amplifiers can be combined to create a differential amplifier that will improve the signal-to-noise ratio. The input leads should be close to each other so that the environmental noise will be the same in both leads. Ideally the two leads will carry the same noise, but only the active lead will carry the signal. A measure of how identical the two amplifiers are is the common mode rejection ratio (CMRR), which is the ratio between the gain of neurophysiological potential and the gain of 60-cycle noise.

# Electrodes

Three electrodes are always attached to the patient during nerve conduction studies (NCS) and EMG; an active electrode, a reference electrode, and a ground electrode.

- The recording electrode (G1) is shown as a black solid circle.
- The reference electrode (G2) is shown as a red or white open circle.
- Ground electrodes are placed between the stimulating and recording points.

Disposable needle electrodes are popular because of their lower maintenance and because of concerns about infection. Reusable needle electrodes have the disadvantage of bending, and of becoming corroded, barbed, or excessively thin, and possible fraying of the Teflon coating, with increased risk of artifacts.

# Needle EMG Electrodes

(a) Surface electrodes. These electrodes come in different sizes and forms and are usually applied with adhesive to the skin over the muscle under study. To reduce impedance, the skin may be cleansed with alcohol and the surface scraped if needed. Surface electrodes are suitable for recording summated muscle action potentials. They are primarily used for recording compound nerve or muscle action potentials or for stimulating the peripheral nerves, as a ground lead, and as a reference with monopolar needle electrodes.

(b) Monopolar needle electrodes. These electrodes are thin needles with a fine point, insulated except for the distal 0.2–0.4 mm. Their average diameter is 0.8 mm and they are insulated with a sleeve of Teflon. Monopolar needle electrodes produce larger amplitudes and greater phasicity and are less painful and less expensive when compared with concentric needle electrodes. A disadvantage is that they are less stable electrically and, hence, they are noisier. They register the potential/voltage difference between the tip and surface. Pick up is 360°.

(c) Concentric (coaxial) needle electrodes. These consist of a central wire (G1) covered with insulation, and the surface of the needle is G2. An advantage is that the recorded potentials are sharper than those with monopolar needle electrodes because of the better rise time, with a clear signal and less noise. A disadvantage is that, compared with monopolar needle electrodes, it they are usually larger in diameter and hence more painful, and they are also more expensive. Pick up is 180°. They register the potential/voltage difference between the wire and shaft.

(d) Single-fiber electrode. The single-fiber electrode is a 25-μm-diameter electrode mounted on the side of the needle. The single-fiber needles may contain two or more wires exposed along the shaft, serving as a leading edge. This electrode enables one to achieve selective recordings of discharge from single muscle fibers rather than from single motor units.

(e) Multi electrodes. These flexible wire electrodes and glass microelectrodes are not for clinical use.

# Stimulus and Stimulators

In clinical NCS, an electrical stimulus is applied to a nerve. The stimulus may be described with respect to the evoked potential of the nerve or muscle. The threshold is one that is just sufficient to produce a detectable response.

- Subthreshold stimulation: Any stimulation that is less than the threshold stimulation.
- Maximal stimulation: Any stimulus intensity after which a further increase in intensity causes no increase in the amplitude of the evoked potential.
- Submaximal stimulation: Stimulation of intensity below the maximal level but above the threshold is called sub-maximal stimulation.
- Supramaximal stimulation: Stimulation of an intensity greater than the maximal stimulus is termed supramaximal stimulation. By convention, supramaximal stimuli are used for NCS. Electrical stimulation of approximately 20 % greater than that required for the maximal stimulation is used for supramaximal stimulation.

At low intensities of stimulation, the electric current will flow in the superficial soft tissues. As the intensity of the stimulus increases, more of the current will enter the axon at the cathode. Stimulators are used to activate action potentials in the nerve fiber. The flow of current is from the anode (+ pole) of the stimulator to the cathode (− pole). Standard stimulation techniques require the poles to be 2–3 cm apart. By convention the stimulating electrodes are called:

> Bipolar if they are encased or attached together
> Monopolar if they are not encased and not together
> The cathode is shown as a black solid circle.
> The anode is shown as a red or white open circle.

Upon stimulation, the initial depolarization allows sodium ions to flow in, and an action potential is generated under the cathode. This action potential is capable of propagating in both directions of the axon. However, the part under the anode is hyperpolarized, so the action potential may fail to propagate past the anode (anodal block).

## Safety Factors

Small "leakage currents" can flow to the machine chassis. This is potentially dangerous should the grounded patient touch the chassis. This potentially dangerous situation is prevented by using the safety ground that is electrically connected to the instrument chassis or any electrical equipment that the patient encounters, such as an electric bed. It is important to ensure that the chassis ground wire is not disconnected, and any electrical devices such as an electric bed are grounded.

A leakage current is one that leaks to the instrument, hardware, or recording electrodes that can be released as an electric shock when contacted by the patient or the electrodiagnostic medicine consultant. A three-pronged power cord provides grounding and helps solve the problem of current leakage. Common problems that could result in loss of ground or excessive leakage of the current include faulty wires, the use of two-pin extension cords, and fluid spills. The electrodiagnostic instruments should have annual safety inspections performed by a qualified biomedical equipment technician.

Major health risks of NCS include transmitting infectious diseases and causing bleeding. Electrodiagnostic studies can be performed on patients with implanted cardiac pacemakers, but with caution. Typically, the closer the stimulation site is to the pacemaker and pacing leads, the greater the chance that the voltage will be of sufficient amplitude to inhibit the pacemaker. Hence, it is particularly important that patients with cardiac pacemakers be properly grounded.

Nerve conduction studies are not suggested for patients with an external electrically conductive lead terminating in or near the heart. Individuals who are susceptible to electrical injury are the frail elderly and those in whom the body's natural barriers have been compromised, such as with the use of central venous or arterial catheters. Stimulation should be avoided close to areas with percutaneous catheters.

# Effect of Temperature on Nerve Conduction

Most laboratories maintain the temperature of the upper extremities between 34 °C and 36 °C and that of the lower extremities between 32 and 34 °C.

## *Temperature Normalization Methodologies*

1. Immersion of the limb in a hot water bath.
2. Placing the limb in an infrared heating apparatus.
3. Application of ice or cold water to normalize high temperatures.
4. Use of electric/heating or cooling blankets.
5. Changing the temperature with chemical hot packs, reusable microwave packs, hair dryers, etc. is unreliable.

## *Correction Factors*

Henriksen [2] described a reduction of 2.4 m/s per degree Celsius between 38 and 39 °C for peripheral nerves in the upper extremities. Also, Gasser and Trojaborg [1] recommend a correction factor of 1.7–1.9 m/s per degree Celsius for the near nerve temperature of the sciatic nerve.

## *Practical Points*

- Cooling reduces the maximal force generated by a muscle contraction.
- Endurance time is known to be longer at cooler temperatures than at warmer temperatures.

- After brief exercise, the compound muscle action potential increases in amplitude (pseudofacilitation).
- Cool temperatures tend to enhance neuromuscular junction transmission.
- Jitter is expected to decrease with higher temperatures and increase with cooler temperatures.
- Increasing temperature in a person with demyelinating disease (for example, multiple sclerosis) may increase the symptoms, causing previously "silent" lesions to become manifest (Uhthoff's phenomenon).
- In single-fiber EMG, there will be increased blocking of impulses in patients with myasthenia gravis.

When a motor axon is depolarized the action potentials travel distally and excite the muscle fiber more or less at the same time. The variation in the time interval between the two action potentials of the same motor unit is called as "jitter". SFEMG measures the variation of this inter potential interval (jitter).

# Further Reading

Gasser M, Trojaborg W. Clinical and electrophysiological study of pattern of conduction times in distribution of sciatic nerve. J Neurol Neurosurg Psychiatry. 1964;27:351–7.

Henriksen JD. Conduction velocity of motor nerves in normal subjects and patients with neuromuscular disorders. Minneapolis: University of Minnesota; 1956.

Leporefe FE. Uhthoff's symptom in disorders of the anterior visual pathways. Neurology. 1994;44:1036–8.

# Chapter 7
# Nerve Conduction Studies

Nerve conduction studies evaluate neuromuscular disease by providing a neurophysiological assessment of peripheral nerves, neuromuscular junctions, muscles, dorsal root ganglion cells, and anterior horn cells. The peripheral nerve is studied with regard to its pathophysiology and its localization.

Motor nerve conduction studies assess the motor axons by selectively recording the muscle response to nerve stimulation. Similarly, sensory nerve conduction studies access the sensory axons by stimulating and recording from the peripheral nerves or the peripheral sensory axons.

The value of nerve conduction studies is their capacity to assess the neurophysiology of the lower motor neuron structure and to localize abnormalities along the path of a nerve. These studies can then characterize whether a neuropathy arises from abnormalities of the motor nerve, sensory nerve, or mixed nerve, and whether the disorder is a demyelinating or an axonal one.

The anterior horn cells, neuromuscular junctions, and the muscles can also be studied. Anomalous innervation is also identified. These studies provide objective measurements with regard to localization, severity, and pathophysiology, and they can also help with prognosis.

A.Q. Rana et al., *Neurophysiology in Clinical Practice*,
In Clinical Practice, DOI 10.1007/978-3-319-39342-1_7,
© Springer International Publishing Switzerland 2017

Nerve conduction studies evaluate the efficiency and speed at which the nerves can send electrical signals.

- Attention to details of technique.
- Stimulation of nerve over an accessible point/s and recording at a distance (r).
- Performed on any accessible nerve (peripheral or central).
- Latency is the time in milliseconds from the stimulus to the recorded response.
- NCV (M/cc) equals the distance between the SNR electrode and the oblique delta sign of latency [1, 2].
- Information obtained in nerve conduction studies:

  (a) Conduction characteristics
  (b) Analysis of action potential morphology (shape, amplitude, duration, and more)

## Near Nerve Stimulation

Needle electrodes may be used for nerve stimulation, but there is a risk of hydrolysis or ionization occurring at the tip of the needle electrode. To minimize this risk, the needle tip should have a larger surface area than the current needs to be distributed over and this can be achieved by stripping back the Teflon insulation to distribute the current over a larger area. Concentric electrodes should not be used because the small area of the central electrode core will lead to a much higher current density.

## Motor Nerve Conduction Studies

- Stimulate motor nerve at an accessible point and record from a muscle innervated by the nerve (compound muscle action potential; CMAP).
- Use supramaximal stimulus (>25 %).

- Latency is the duration between the stimulus and the onset of the initial negative deflection.
- Amplitude (CMAP) baseline-to-peak recording or peak-to-peak recording.
- Orthodromic stimulation is in the direction of physiological conduction with stimulation distally and recording proximally.

## Sensory Nerve Conduction Studies

- Stimulate nerve and record nerve (no muscle, no synapse).
- Orthodromic conduction study stimulates distally and records proximally.
- Antidromic conduction study stimulates proximally and records distally; opposite to physiological conduction.
- Measure baseline to negative peak or peak-to-peak.
- Signal averages are useful to record low-amplitude potentials.
- Use a supramaximal stimulus of more than 25 %; avoid contamination with CMAP.

## Factors Influencing Conduction Velocity

1. Age. Nerve conduction velocity in term infants is approximately 50 % of the adult values. Adult values are reached by age 5 years; conduction slowing of about 10% occurs at 60 years of age.
2. Site of recording. Proximal conductions are faster than conductions in the distal segments of the limbs.
3. Fiber type. Large-diameter fibers have faster conduction velocity, small-diameter myelinated fibers have slower conduction, and unmyelinated fibers have even slower conduction. The typical internodal distance is 1 mm, and internodal conduction is 20 µm/s. Hence, conduction velocity typically is 50 m/s.

4. Temperature. The relationship between temperature and conduction velocity is mostly linear from 29 to 38 C. Conduction velocity increases by approximately 5% per degree centigrade over this temperature range. Hence, a change of 2–3 m/s occurs per degree in normal nerve conduction at 40–60 m/s [3].
5. Focal compression. Conduction also slows in the absence of demyelination if the fiber diameter is decreased (reduced fiber diameter = decreased capacitance = decreased conduction).

# References

1  Thomas JEN, Lambert EH. Ulnar nerve conduction velocity and H-reflex in inference in children. J Appl Physiol. 1960;15:1.
2  Wagman IH, Lesse H. Maximum conduction velocities of motor fibers of ulnar nerve in human subjects of various ages and sizes. J Neurophysiol. 1952;15:235.
3  Hendriksen JD. Conduction velocity of motor nerves in normal subjects and the patients with neuromuscular disorders. Thesis, University of Minnesota, Minneapolis, 1966. The fourth reference is Johnson EW, Olson KJ. Clinical value of motor nerve conduction velocity determination. JAMA. 1960;172:2030.

# Chapter 8
## Sensory Peripheral Neuropathy

Sensory nerve conduction studies are an essential part of the electrodiagnostic examination, for several reasons. Some peripheral nerve lesions involve only sensory nerves or only the sensory fibers of mixed nerves. Sensory conduction studies are typically more sensitive than their motor counterparts to pathophysiological processes occurring along mixed nerves. For example, in these mixed nerves, sensory conduction typically occurs before or is more severe than the conduction in the motor components, such as in the segmental demyelinating lesions of carpal tunnel syndrome. Sensory conduction amplitudes are more sensitive to axon loss than motor nerves.

Sensory conduction amplitudes are not affected by lesions that are proximal to the dorsal root ganglion, such as the intraspinal lesions of myelopathy or radiculopathy. Because the cutaneous nerves are more superficial than the motor nerve fibers, they are more susceptible to trauma, and consequently the sensory responses may be absent or of low amplitude because of coincidental nerve injury.

In the lower extremities, attention should be paid to prior surgical procedures such as vein stripping, phlebotomies, or tendon-lengthening procedures that result in an absence of sensory nerve action potentials (SNAPs). The SNAPs are also affected by lymphedema; gross obesity; and thickened, hyperkeratotic, or calloused skin. Lower extremity sensory conduction studies show abnormalities much sooner than upper

A.Q. Rana et al., *Neurophysiology in Clinical Practice*,
In Clinical Practice, DOI 10.1007/978-3-319-39342-1_8,
© Springer International Publishing Switzerland 2017

extremity studies, such as in patients with generalized peripheral polyneuropathies.

Side-to-side comparison of the sensory responses can be helpful when evaluating unilateral abnormalities. A response of 50 % or lower than the response obtained on the asymptomatic side is considered abnormal in many laboratories. Absent sensory responses can occur as a result of age.

There are several differences in the parameters reported by electromyography laboratories. These include having variable distances between stimulating and recording electrodes, reporting conduction times as latencies versus conduction velocities, and measuring latencies from onset to peak or measuring from peak-to-peak versus baseline-to-peak.

# Technical Considerations in Sensory Nerve Conduction Studies

- Stimulate nerve and record nerve (no muscle, no synapse).
- The requirement in each conduction study is the recording of an evoked potential.
- Orthodromic conduction stimulates distally and records proximally.
- Antidromic conduction stimulates proximally and records distally; opposite to physiological conduction.
- Measure baseline to negative peak or peak-to-peak.
- Signal averaging is useful to record low-amplitude potentials.
- Use a supramaximal stimulus of more than 25 %, but avoid contamination with compound muscle action potential (CMAP).
- The ground is optimally placed between the stimulating and the recording electrodes.

## Sural Nerve

The sural nerve is derived from the S1 and S2 roots. It is formed by a medial sural nerve, which is a branch of the tibial nerve given off at the inferior angle of the popliteal fossa.

This branch of the tibial nerve, along with the communicating branch of the peroneal nerve, the lateral sural nerve, forms the sural nerve. The sural nerve leaves the popliteal fossa in the groove between the two heads of the gastrocnemius, and it becomes superficial at approximately the junction of the middle and lower third of the leg. It then continues, to pass behind the lateral malleolus and the lateral border of the foot.

**Stimulation**
Antidromic surface stimulation of the nerve is distal to the lower border of the bellies of the gastrocnemius, at the junction of the middle and the lower third of the leg. This is at a distance of 10–16 cm above the lateral malleolus.

**Recording**
The active electrode is placed between the lateral malleolus and the Achilles tendon.

**Values**
Mean latency is 3.8 ms and amplitudes vary between 10 and 75 μV.

# Superficial Peroneal Nerve

The superficial peroneal nerve is a branch of the common peroneal nerve that is formed from the L4, L5, and the S1 nerve roots. It provides cutaneous innervation to the distal lateral leg and most of the dorsum of the foot. The nerve becomes superficial in the groove between the peroneus longus and the extensor digitorum longus at the junction of the middle and distal thirds of the leg. The nerve passes in front of the extensor retinaculum at the ankle to reach the dorsum of the foot.

**Stimulation**
Antidromic surface stimulation is performed 10–15 cm proximal to the lateral malleolus and anterior to the peroneus longus.

**Recording**

The active surface electrode is placed above the junction of the lateral third of a line connecting the malleoli.

**Values**

Mean latency is 3.2 ms and amplitudes are 10–70 μV. The amplitude is generally half that of the sural nerve.

# Saphenous Nerve

The saphenous nerve is the longest and the largest branch of the femoral nerve. It provides innervation to the anteromedial and posteromedial leg and the medial border of the foot to the base of the first toe. The origin is from the L3-L4 roots. The saphenous nerve branches from the femoral nerve approximately 4 cm distal to the inguinal ligament and travels deep to the sartorius muscle in the adductor canal. It remains anterior to the femoral artery and crosses from the lateral to the medial position in the middle third of the thigh. The nerve then continues along the medial condyle of the femur, where along appears as the fascia and travels between the sartorius and gracilis tendons up to the medial malleolus.

### Stimulation

Antidromic surface stimulation is performed over the medial aspect of a slightly flexed knee at a point between the sartorius and gracilis tendons approximately 1 cm above the lower pole of the patella.

### Recording

Surface recording is from a line drawn from the stimulation point directly 15 cm to the medial border of the tibia.

### Values

Latency is 2.5 ms and amplitude is 7–15 μV.

# Lateral Femoral Cutaneous Nerve

The lateral femoral cutaneous nerve is formed from the dorsal portions of the ventral primary divisions of the L2, L3 roots and is a purely sensory nerve. It has an intraabdominal course, and it emerges from the lateral border of the psoas muscle and travels along the iliacus to the anterior superior iliac spine (ASIS). It emerges under the later portion of the inguinal ligament and travels subcutaneously through the origin of the sartorius. The anterior branch provides innervation to the anterolateral thigh and as distally as to the knee. The posterior branch innervates the skin of the lower lateral quadrant of the buttock to the mid thigh.

### Stimulation
Antidromic surface stimulation is performed 1 cm medial to the ASIS at the inguinal ligament over the origin of the sartorius muscle.

### Recording
Surface electrodes are placed along a line connecting the ASIS to the lateral border of the patella. The active electrode is placed 17–20 cm distal to the ASIS.

### Values
Latency is 2.8 ms, and SNAPs are generally 4–11 μV.

# Superficial Radial Nerve

The superficial radial nerve is the terminal sensory division of the radial nerve derived from the C6 and C7 roots. It supplies the dorsal skin of the lateral two-thirds of the hand, the lateral two and a half digits, and the ball of the thumb. Its course remains deep through the forearm until the lower third, where it emerges dorsally near the brachioradialis tendon, and its branches run over the tendons of the thumb extensors.

**Stimulation**
Surface stimulation is performed on the dorsolateral aspect of the radius approximately 7 cm above the radial styloid process. The stimulating cathode should be 10 cm proximal to the recording electrode.

**Recording**
The surface electrode is placed over the extensor pollicis longus tendon at the wrist.

**Values**
Latency is approximately 1.6 ms, and peak-to-peak amplitude is 16–86 μV.

# Dorsal Cutaneous Branch of the Ulnar Nerve

The dorsal cutaneous branch of the ulnar nerve is the major sensory nerve arising from the distal third of the forearm and it carries fibers from the C8, T1 roots via the medial cord of the brachial plexus. This branch innervates the dorsal skin over the medial metacarpal region and the median digits. The dorsal cutaneous nerve becomes superficial approximately 5 cm above the wrist, where it lies between the flexor carpi ulnaris tendon and the ulnar bone.

**Stimulation**
With the arm supinated, the nerve is stimulated 5 cm above the ulnar styloid between the flexor carpi ulnaris tendon in the ulnar bone.

**Recording**
The recordings of this electrode are placed 8 cm from the stimulating site at the apex between the fourth and the fifth metacarpal bones.

**Values**
Latency is 2.0, and the amplitude is 20 μV.

# Lateral Antebrachial Cutaneous Nerve

The lateral antebrachial cutaneous nerve is the sensory continuation of the musculocutaneous nerve. The fibers are derived from the C5 and C6 roots. The fibers travel through the lateral cord of the brachial plexus and provide sensory innervation from the lateral volar forearm to the wrist. A dorsal branch supplies the lateral dorsal aspect of the forearm, again, up to the wrist.

### Stimulation
Antidromic surface stimulation is applied at the level of the elbow lateral to the biceps brachii tendon.

### Recording
Recording is done from a surface electrode placed along a line connecting the stimulation point to the radial artery, or to the radial styloid process at the wrist. The active recording electrode is placed 12 cm distal to the stimulating cathode.

### Values
Latency is 1.8 ms and SNAP is between 15 and 50 µV.

# Medial Antebrachial Cutaneous Nerve

The medial antebrachial cutaneous nerve is derived from the C8, T1 root and courses along the medial cord of the brachial plexus. This is a pure sensory nerve innervating the anterior upper arm and the medial forearm up to the wrist.

### Stimulation
Surface stimulation is performed medial to the brachial artery 4–5 cm proximal to the medial humeral epicondyle.

**Recording**
The surface electrode is placed along a line connecting the point halfway between the medial epicondyle and the biceps brachii tendon drawn to the ulnar styloid process. The active electrode is 7–8 cm distal to the epicondyle.

**Values**
Mean latency is 1.9 ms and amplitudes vary between 8 and 4 μV.

# Posterior Antebrachial Cutaneous Nerve

The posterior antebrachial cutaneous nerve is a sensory branch of the radial nerve in the spiral groove. It derives fibers from the C5 to the C8 roots and traverses the posterior cord of the brachial plexus at the radial nerve. It provides sensations to the skin and the lateral arm and the elbow and the dorsal forearm distally up to the wrist. It originates along lateral head of the triceps. It innervates the skin of the distal upper arm, lateral upper arm, and the dorsum of the forearm.

**Stimulation**
Surface stimulation is performed at the elbow just above the lateral epicondyle between the biceps brachii and the triceps brachii.

**Recording**
The surface electrode is placed along a line that extends from the stimulation point to the mid dorsum of the wrist, midway between the ulnar and the radial styloid processes. The active electrode is placed approximately 12 cm distal to the stimulating cathode.

**Values**
Latency is 1.9 ms and amplitude varies between 5 and 20 μV.

# Median Digital Nerves

Distal to the carpal tunnel, the median nerve divides into sensory and motor branches. The digital sensory branches supply the index finger, the middle finger, and part of the thumb and fourth finger.

### Stimulation
Orthodromic stimulation can be undertaken by placing ring electrodes around the second and the third digits near the metacarpophalangeal joints.

### Recording
The recording electrode is placed over the median nerve on the anterior aspect of the wrist 13 cm proximal to the ring cathode. Antidromic studies can also be undertaken by stimulating the median nerve at the mid wrist with recordings from ring electrodes placed around the second and the third digits, near the metacarpophalangeal joints.

### Values
Normal sensory orthodromic latencies are 3.0 ms, and antidromic latencies are 3.2 ms. Amplitudes are more than 15 µV orthodromically, and with antidromic stimulation the amplitudes are equal to or slightly higher than the orthodromic.

# Ulnar Digital Nerve

The sensory branch arises in the wrist proximal to Guyon's canal.

### Stimulation
Antidromic stimulation is undertaken by placing ring electrodes around the fourth and the fifth digits near the metacarpophalangeal joint.

**Recording**
Recording is done from the ulnar nerve and the wrist at a distance of 13 cm from the ring electrode. Antidromic stimulation can also be undertaken by stimulating the ulnar nerve at the wrist and recording at a 13-cm distance over the fourth and fifth digits near the metacarpophalangeal joints, by placing ring electrodes.

**Values**
Sensory latencies in orthodromic stimulation are generally 2.8 ms, and the amplitude is generally more than 8 μV. In antidromic stimulation, the sensory latency is 3.2 ms, and the amplitudes are the same or slightly higher than the orthodromic.

## Radial Digital Nerve

The radial nerve separates from the deep motor branch near the elbow. Also, in the distal forearm, the radial sensory nerve becomes superficial, crossing the radius. This superficial radial nerve supplies the lateral aspect of the dorsum of the hand and the proximal dorsum of the first three and a half digits. The digital branch supplies the radial or lateral aspect of the thumb.

TABLE 8.1  Normal values for nerve conduction studies (NCS) in our laboratory

| Nerve | Distal latency | Nerve conduction velocity | Amplitude |
|---|---|---|---|
| Sensory NCS *Upper limb* | | | |
| Median | ≤3.4 ms @ 13 cm | ≥50 m/s | ≥8 μV |
| Ulnar | ≤3.2 ms @ 11 cm | ≥50 m/s | ≥8 μV |
| Radial | ≤2.8 ms @ 10 cm | ≥50 m/s – | ≥8 μV |
| *Lower limb* | | | |
| Sural | ≤4.4 ms @ 14 cm | ≥40 m/s | ≥5 μV below age 50 |

TABLE 8.1  (continued)

| Nerve | Distal latency | Nerve conduction velocity | Amplitude |
|---|---|---|---|
| Motor NCS *Upper limb* | | | |
| Median | ≤3.9 ms @ 7 cm | ≥45 m/s | ≥5 mV |
| Ulnar, below elbow | ≤3.5 ms @ 7 cm | ≥45 m/s | ≥6 mV |
| Ulnar, across elbow | – | ≥45 m/s | ≥6 mV |
| Radial | ≤3.4 ms @ 6 cm | ≥45 m/s | ≥5 mV |
| *Lower limb* | | | |
| Peroneal | ≤5.0 ms @ 8 cm | ≥40 m/s | ≥3 mV |
| Tibial | ≤5.0 ms @ 10 cm | ≥40 m/s | ≥5 mV |

TABLE 8.2  Normal values for F-wave latencies in our laboratory

| Nerve | Latency |
|---|---|
| *Upper limb* | |
| Median | ≤31 ms |
| Ulnar | ≤31 ms |
| *Lower limb* | |
| Peroneal | ≤51 ms |
| Tibial | ≤55 ms |

F-wave latency may vary with height

# Chapter 9
# Nerve Trauma

## Seddon's Classification of Nerve Injuries [1]

1. *Neurapraxia*

   Neurapraxia is a temporary loss of function without discontinuity of the axon. The mildest form arises from the injection of a local anesthetic or a transient loss of circulation, such as with leg crossing. The changes are reversible. The reversibility can occur within a few weeks to a few months. Conduction velocity may be reduced, and focal demyelination may occur. In demyelinating neuropathies, associated neurapraxia is a major cause of paralysis.

2. *Axonotmesis*

   Axonotmesis is a loss of continuity of the axons and their myelin sheaths, with preservation of the connective tissue. It usually occurs 4 or 5 days after an acute disruption. The distal segment becomes inexcitable as a result of nerve fiber degeneration (Wallerian degeneration). A complete conduction block may occur, such as in neurapraxia. Conduction velocity may remain normal in partial axonotmesis, as the intact axons conduct normally. In a motor nerve, the compound muscle action potential would be reduced in amplitude in proportion to the amount of axonal loss. Electromyography would show evidence of axonal loss in the form of fibrillation potentials and sharp positive waves 2–3 weeks after the occurrence of axonotmesis. Axons regenerate at a rate of approximately 1–3 mm per day.

A.Q. Rana et al., *Neurophysiology in Clinical Practice*,
In Clinical Practice, DOI 10.1007/978-3-319-39342-1_9,
© Springer International Publishing Switzerland 2017

3. *Neurotmesis*
   Neurotmesis is the complete or almost complete severance or disorganization of the nerve. Sunderland [2] described three subdivisions of neurotmesis.

   (a) The perineurium and the architecture of the nerve sheath are preserved. Regeneration may occur, but less completely than from an axonotmesis. There is increased risk of misdirected innervation, leading to synkinesis.

   (b) Here the perineurium is also damaged but continuity of the nerve remains grossly intact. Some regeneration may occur.

   (c) There is complete separation of the nerve, with a loss of continuity. Surgical repair is the only remaining option.

# Schaumberg's Modification of Seddon's Classification [3]

## Class I Injuries

Class I injuries (neurapraxia) result in a reversible blockade of nerve conduction, and this is related to mild or moderate focal compression. Class I injuries may result in decreased strength, an absence of deep tendon reflexes, and loss of sensations. There is usually no loss of autonomic nerve function. There is no permanent damage to the axon. Recovery is spontaneous and occurs within 3 months. Class I injuries are further divided into two types:

**Type I**  Mild and rapidly reversible injury resulting from transient ischemia, with no anatomical demyelinating changes

**Type II**  Slowly reversible conduction blockade, resulting from paranodal demyelination

## *Class II Injuries*

Class II injuries (axonotmesis) result in variable loss of motor, sensory, and sympathetic nerve function. Myelinated and unmyelinated fibers may both be involved in this process. Muscle atrophy and areflexia are consistently found. Axonal damage is evident, but the Schwann cell remains intact, along with the endoneurium. Regeneration is generally effective, with a high degree of success because of the integrity of the Schwann cell basal lamina and endoneurium. Recovery is slow because axonal regeneration occurs at the rate of 1–8 mm per day [4]. Prognosis is good because axons regenerate in the uninterrupted connective tissue scaffolding.

## *Class III Injuries*

Class III injuries (neurotmesis) are commonly the result of complete nerve transection, such as that resulting from a stab wound; these injuries can also arise from nerve traction injury that also disrupts the connective tissue components, and the nerve trunk is completely transected. Wallerian degeneration takes place in the region distal to the lesion. Regeneration may occur, but with little success. Prognosis is poor for the restoration of end-organ function.

# References

1. Seddon HJ. Three types of nerve injury. Brain. 1943;66:237.
2. Sunderland S. Nerve and nerve injuries, Edition 2. Edinburgh: Churchill Livingston; 1978.
3. Schaumberg HH, Spencer PS, Thomas PK. Disorder of peripheral nerves. Philadelphia: FA Davis; 1983.
4. Seltzer ME. Regeneration of peripheral nerve. In: Sumner AJ, editor. The physiology of peripheral nerve disease. Philadelphia: WB Saunders; 1980.

# Chapter 10
## Peripheral Neuropathy

The term "peripheral neuropathy" covers a variety of clinical syndromes that affect numerous types of nerve cells and fibers, including motor, sensory, and autonomic fibers. Nearly all peripheral neuropathies affect all these fiber types to a certain degree. However, a single fiber type may be targeted exclusively, and may be affected most, to a greater extent in some disorders than others. There are four cardinal patterns of peripheral neuropathy, including polyneuropathy; a few clinically important polyneuropathies are described below:

1. Polyneuropathy
2. Mononeuropathy
3. Mononeuritis multiplex
4. Autonomic neuropathy

## Classification of Polyneuropathies

There are several ways to classify polyneuropathies. This classification depends on the following:

- Course or progression (acute, subacute, chronic, progressive, relapsing)
- Fiber type (motor, sensory, small fiber, large fiber, autonomic)
- Pattern of involvement (symmetrical, multifocal)

A.Q. Rana et al., *Neurophysiology in Clinical Practice*,
In Clinical Practice, DOI 10.1007/978-3-319-39342-1_10,
© Springer International Publishing Switzerland 2017

- Underlying pathology (autoimmune, compression, ischemia)
- Part of the nerve cell mainly affected (axon, myelin sheath, cell body)
- Hereditary nature (Charcot-Marie-Tooth disease, hereditary neuropathy with a propensity to experience pressure palsy)
- Associated illnesses (porphyria, hypervitaminosis, human immunodeficiency virus)
- Exposure to toxins (occupational, chemotherapy, poisoning)

## Parts of the Nerve Cell Affected

*Axonopathy* is a disorder that predominantly affects the axons associated with the peripheral nerve fibers. This disorder may be provoked by various metabolic diseases, including diabetes, malnutrition, renal failure, and alcoholism; diseases of the connective tissues; or by effects related to drugs/toxins such as chemotherapy agents. Large-fiber, small-fiber, or both types of axons are affected. The most distal portions of axons are usually the first to degenerate (this is also known as "dying-back neuropathy"), and axonal atrophy is centripetal.

*Myelinopathy*, also referred to as "demyelinating polyneuropathy," is a result of a loss of myelin, leading to conduction slowing or blocking of action potentials through the axon of the nerve cell. The most typical cause of this condition is acute inflammatory demyelinating polyneuropathy. Other causes include nerve entrapments, genetic metabolic disorders, and toxins.

*Neuronopathy* is the result of a derangement of neurons. Motor neuron diseases, toxins, autonomic dysfunction, and sensory neuronopathies may all be a cause of this. Vincristine, a chemotherapy agent, is just one of many neurotoxins that may also cause neuronopathies.

Although polyneuropathies are often suggested by physical examination and history alone, electrophysiological testing is still a large part of the diagnosis, evaluation, and classification of these diseases. A few clinically important polyneuropathies are described below.

## Chronic Polyneuropathy

Chronic polyneuropathy is a diagnosis that is very common, having an estimated prevalence of 2.4–8 per 100,000. Polyneuropathy itself has been linked to a variety of causes. However, a cause cannot be established for about 10–15 % of patients, even after a thorough evaluation has been conducted. Those individuals who do not have a cause associated with the neuropathy mostly have axonal neuropathy. Recently, this specific type of neuropathy has been commonly referred to as chronic idiopathic axonal polyneuropathy (CIAP). Patients who are affected by CIAP show, in middle to old age, mild sensory and motor symptoms, along with slow deterioration. However, the occurrence of serious disability is not evident.

# Distal Symmetrical Polyneuropathy (Protocol for Evaluation)

A simplified version of a nerve conduction study (NCS) protocol may be used to define the existence of distal symmetrical polyneuropathy:

1. Both sural sensory and peroneal motor NCS are conducted in one lower extremity. If no signs of distal symmetrical polyneuropathy are evident, it is concluded that both studies are normal. If this is the case, then the performance of any further NCS is unnecessary.
2. If the sural sensory or peroneal motor NCS turns out to be abnormal, then it is recommended to conduct further NCS protocols. This includes performing NCS of the ulnar sensory nerves, along with the median sensory and ulnar motor nerves in an upper extremity. NCS may also be conducted for a contralateral sural sensory and one tibial motor nerve. However, proceed with caution when interpreting and

understanding median and ulnar studies, since there is a possibility of superadded compression neuropathy.
3. If a response is not present for any of the given nerves studied (sensory or motor), then NCS of the contralateral nerve should be completed.
4. If there is no presence of a peroneal motor response, then NCS of an ipsilateral tibial motor nerve should be completed.

# Demyelinating Neuropathies

There are several demyelinating diseases associated with the peripheral nervous system. These include:

- Guillain-Barré syndrome and chronic inflammatory demyelinating polyneuropathy
- Anti-MAG (myelin-associated glycoprotein) peripheral neuropathy
- Charcot-Marie-Tooth Disease
- Copper deficiency

## *Guillain-Barre Syndrome (GBS): Key Subtypes*

### Acute Inflammatory Demyelinating Polyradiculoneuropathy (AIDP)

AIDP is an autoimmune disorder that is mediated by antibodies. It is elicited by the presence of antecedent viral or bacterial infections. Demyelination is evident via the analysis of electrophysiological findings. More specifically, inflammatory demyelination may also be associated with the loss of axonal nerves. Nevertheless, remyelination can occur once the immune reaction has been brought to a halt.

### Acute Motor Axonal Neuropathy (AMAN)

AMAN is a pure motor axonopathy. Approximately 67% of patients who are affected are shown to be seropositive for campylobacteriosis. The electrophysiological studies conducted are normal in sensory nerves, but show reduced or even absent conduction in motor nerves. However, the recovery process for those affected by AMAN is usually very quick. Of note, a large proportion of affected individuals are pediatric patients.

## Acute Motor Sensory Axonal Neuropathy (AMSAN)

AMSAN is characterized by Wallerian-like degeneration that is evident in myelinated sensory and motor fibers. It also shows minimal inflammation and demyelination. AMSAN is analogous to AMAN but differs in that it affects sensory nerves and roots. This subtype of GBS primarily affects the adult population.

## *Miller Fisher Syndrome*

Miller Fisher syndrome, another variant of GBS, is a rare disorder. It is characterized by rapidly evolving ataxia, are-flexia, moderate limb weakness, and ophthalmoplegia. Sensory loss is not typically seen, but impairment of proprioception is evident in this syndrome. Cranial nerves III and VI, spinal ganglia, and peripheral nerves all show demyelination and inflammation. Action potentials of the sensory nerve are reduced or are not present, and the tibial H-reflex is typically absent. Most individuals affected by Miller Fisher syndrome have a good prognosis, as resolution usually takes place within 1–3 months.

## *Acute Pandysautonomic Neuropathy*

This is the rarest of all variants of GBS. Both sympathetic and parasympathetic nervous systems are involved in this specific neuropathy. Cardiovascular involvement is also commonly associated with this neuropathy. Cardiovascular features include tachycardia, postural hypotension, dysrhythmias, and hypertension. Symptoms include anhydrosis, blurred vision, and dry eyes. The recovery process is typically slow and incomplete.

## Pure Sensory Guillain-Barré Syndrome

The literature has also described another form of GBS, which is a pure sensory variant of the syndrome. This variant is characterized by a rapid onset of sensory loss, and areflexia occurring in a symmetrical and widespread pattern.

## Electrodiagnostic Criteria for Demyelinating Peripheral Neuropathies

These criteria concern NCS in which the predominant process that has occurred is demyelination. The criteria must have three of the following four features:

I. *Conduction velocity reduced in two or more motor nerves*:

   (a) <80 % of lower limit of normal (LLN) if amplitude is >80 % of LLN
   (b) <70 % of LLN if amplitude <80 % of LLN

II. *Conduction block or abnormal temporal dispersion in one or more motor nerves*:

   1. *Regions for examination are*:

      (a) Either peroneal nerve between ankle and below fibular head
      (b) Median nerve between wrist and elbow
      (c) Ulnar nerve between wrist and below elbow

   2. *Criteria for partial conduction block*:

      (a) Compound muscle action potential (CMAP) change of <15 % in duration between proximal and distal sites
      (b) A >20 % drop in CMAP negative-peak area, or peak-to-peak amplitude between proximal and distal sites

3. *Criteria for abnormal temporal dispersion and possible conduction block*:

   (a) CMAP duration difference of >15 % between proximal and distal sites
   (b) A >20 % drop in negative-peak duration or peak-to-peak amplitude between proximal and distal sites

III. *Prolonged distal latencies in two or more nerves*:

   (a) >125 % of upper limit of normal (ULN) if amplitude >80 % of LLN
   (b) >150 % of ULN if amplitude <80 % of LLN

IV. *Absent F-waves or prolonged minimum F-wave latencies in two or more motor nerves*:

   (a) >120 % of ULN if amplitude >80 % of LLN
   (b) >150 % of ULN if amplitude <80 % of LLN

## *Prognosis for Poor Functional Outcome in Demyelinating Neuropathies*

- Low mean CMAP amplitudes of less than 20 % of the lower limit of normal.
- The existence of inexcitable nerves on initial electrophysiological studies.
- Continuance of CMAP with a low mean on later testing (>1 month after the onset) causes an even greater sensitivity and specificity, compared with results of initial testing following onset.

## *Autoimmune Neuropathy*

This is a class of peripheral polyneuropathy generally associated with antibodies to ganglioside subtypes. Anti-GM1 antibody is frequently found in GBS, motor neuron disease, multifocal motor neuronopathy, and AMAN, a subtype of GBS. Anti-GM1 antibody is also found, at elevated levels, in patients with multiple sclerosis, celiac disease, or infection with *Campylobacter jejuni.*

## Multifocal Motor Neuropathy (MMN)

Multifocal motor neuropathy is a well-defined purely motor polyneuropathy. This type of neuropathy is defined by the existence of multifocal partial motor conduction blocks (CBs) in various nerves. It is also characterized by recurrent association with anti-GM1 and IgM antibodies. Also, there is typically an adequate response to high-dose intravenous immunoglobin (IVIg) therapy.

Clinical presentation is of indolent progression of focal weakness, confined to specific muscles, and later atrophy. Fasciculations and cramps occur occasionally, and there are no sensory symptoms. Furthermore, there is nerve hypertrophy, spinal segmental distribution is unusual, and cranial nerve and respiratory muscle involvement is rare. There is an absence of deep tendon reflexes, and there is a predilection for MMN to affect the arms. Motor NCS may permit identification and localization of the segments of the nerve affected by the underlying pathology.

## Electrodiagnostic Features

· 1. Conduction block, temporal dispersion, or conduction slowing confined to motor axons
2. Fasciculations and occasional myokymia
3. Needle electromyography (EMG) shows chronic and active denervation

# Criteria for the Diagnosis of Multifocal Motor Neuropathy

## Criteria for Definite MMN

There are specific criteria to consider when analyzing for definite MMN. First of all, there should be the presence of weakness without the existence of objective sensory loss, as well as temporal dispersion in two or more peripheral nerves. Additionally, physical and historical data of diffuse symmetrical weakness does not include MMN during the initial stages of symptomatic weakness.

Another criterion to acknowledge involves definite CB. Definite CB should exist in two or more nerves, extrinsic to the common entrapment sites, including the ulnar nerve at the elbow or wrist, the median nerve at the wrist, and the peroneal nerve at the fibular head.

An additional criterion involves velocity; across the same segments with demonstrated motor CB, sensory nerve conduction velocity should be normal.

The final criterion for definite MMN is that, for sensory nerve conduction, normal results must be evident in all the nerves that are tested. Additionally, a minimum of at least three nerves should be tested. Finally, there should be no appearance of the following upper motor neuron signs: clonus, spastic tone, extensor plantar response, and pseudobulbar palsy.

## Criteria for Probable MMN

When examining for probable MMN, certain criteria are also available for reference. Firstly, there should be the presence of weakness without the existence of objective sensory loss, as well as temporal dispersion in two or more named nerves.

Additionally, the presence of diffuse symmetrical weakness does not include probable MMN during the initial stages of symptomatic weakness.

Secondly, in two or more motor nerve segments that are not common entrapment sites, there should be the existence of a probable CB. It is also possible that there could be a definite CB in one motor nerve segment and a probable CB in another distinct motor nerve segment; in this scenario, none of these segments is a common entrapment site.

Furthermore, across related segments with a demonstrated motor CB, there should be normal sensory nerve conduction velocity. At the same time, the study of this segment should be technically viable. This means that studies are not mandatory for segments that are proximal to either the axilla or the popliteal fossa.

Also, for a minimum of three nerves tested, normal results should be obtained for sensory NCS in all of the tested nerves.

As stated above, there should also be no appearance of the following upper motor neuron signs: clonus, spastic tone, extensor plantar response, and pseudobulbar palsy.

A CB is defined as a reduction in the amplitude or area of the CMAP in proximal sites. This reduction is evident when compared with distal stimulation; it is also accompanied by no significant or only focal abnormal temporal dispersion. Typically a 50 % drop in CMAP is considered abnormal (60 % drop is considered abnormal for tibial nerves).

## Sensory Neuropathy (Dorsal Root Ganglionopathy)

Sensory neuropathies are less frequent than motor neuropathies. Sensory neuropathies are characterized by abnormal sensory NCS with normal motor conductions, and normal needle EMG studies. The diagnosis requires a thorough examination of at least three limbs. Typically, the sensory responses are diffusely low in amplitude or absent, and in the H-reflexes they are unelicitable.

## *Multiple Mononeuropathies*

The involvement of two or more nerves by the same pathological process is referred to as multiple mononeuropathy or mononeuropathy multiplex. Extensive involvement may produce a clinical picture that would be difficult to differentiate from a generalized polyneuropathy. Diagnosis requires extensive investigation to locate asymmetries in the pattern of abnormality. Multiple mononeuropathy is often due to vasculitis.

## *Small Fiber Neuropathy*

Small fiber neuropathies are a common disorder. They often present with painful feet in patients over the age of 60. Autoimmune mechanisms are often suspected but are rarely identified. Known causes of small fiber neuropathy are diabetes mellitus, toxins, amyloidosis, and inherited sensory and autonomic neuropathies. Small fiber neuropathy can be diffuse or multifocal. It can be subclinical or significantly symptomatic. Diagnosis is based on clinical features, as NCS are often normal. Specialized tests of small fiber function can be conducted. These include the examination of epidermal nerve fiber density, and also sudomotor, quantitative sensory, and cardiovagal testing. A range of 59–88 % is given for the sensitivities of these tests.

# Further Reading

Asbury AK, Cornblath DR. Assessment of current diagnostic criteria for Guillain-Barre syndrome. Ann Neurol. 1990;27(suppl): S21–4.

England JD, Gronseth GS, Franklin G, et al. Distal symmetric polyneuropathy: a definition or clinical research. Neurology. 2005;64:199–207.

Lacomis D. Small-fiber neuropathy. Muscle Nerve. 2002;26: 173–88.

Olney RK, Lewis RA, Putnam TD, Campellone JV. Consensus criteria for the diagnosis of multifocal of motor neuropathy. Muscle Nerve. 2003;27:117–21.

Rosenberg NR, Vermeulen M. Chronic idiopathic axonal polyneuropathy revisited. J Neurol. 2004;251:1128–32.

Taylor PK. CMAP dispersion, amplitude DK and area DK in a normal population. Muscle Nerve. 1993;16:1181–7.

# Chapter 11
# Radiculopathies

Degenerative and traumatic spinal diseases commonly cause lesions of the cervical (CR) and lumbosacral (LR) nerve roots. Less often, vascular and inflammatory disorders can cause non-compressive nerve root lesions.

The most common clinical presentation in a radiculopathy is neck or back pain, arm or leg pain, or extremity paresthesia. Furthermore, there may be increased pain with coughing, sneezing, or the Valsalva maneuver, and reproduction of symptoms with a change in position, particularly for CR. Sensory symptoms consistent with injury to the sensory root fibers are numbness, pain, and paresthesia.

Electrodiagnostic (EDX) studies alone are capable of identifying physiological dysfunction, thus identifying ongoing injury to the nerve roots. This information may not be available through imaging studies. Further imaging studies cannot evaluate inflammatory or vascular nerve root damage.

Electrodiagnostic studies can demonstrate evidence of radiculopathy with both compressive and non-compressive etiologies. Although EDX studies are sensitive and specific for defining nerve root injury, EDX evaluation cannot determine the cause of the injury.

Bilateral studies are required to rule out a central disc herniation with bilateral radiculopathies, or spinal stenosis. The EDX evaluation needs to be extensive enough to

A.Q. Rana et al., *Neurophysiology in Clinical Practice*,
In Clinical Practice, DOI 10.1007/978-3-319-39342-1_11,
© Springer International Publishing Switzerland 2017

develop the differential diagnosis of radiculopathy, plexopathies, polyneuropathies, or mononeuropathies—all of which can present with similar signs and symptoms.

# Needle Electrode Examination

Needle electrode examination (NEE) is the single most useful procedure for the assessment of a nerve root lesion, as this examination assesses the physiological integrity of the root. The advantages of NEE for the evaluation of radiculopathy are that it identifies

- abnormalities in myotome distribution, where multiple muscles innervated by the same spinal nerve root are involved, and these abnormalities can define root injury; NEE also
- identifies the specific level or levels of a root injury and differentiates between root injury and other peripheral nerve lesions.
- Defining the severity and duration of a root injury
- When evaluating subclinical damage to the motor root that is evident with NEE but not evident from clinical manual muscle strength testing,

  important considerations are that:

- The NEE will be abnormal when the degree of the injury is sufficient to produce motor axon loss, conduction block, or both.
- Such injury is not necessarily present in the motor fibers to all the muscles of a particular myotome.
- To adequately screen each major myotome in the symptomatic limb, NEE of a sufficient number of muscles (five to seven), including the paraspinal muscles, is necessary.
- Myotomal maps or charts may be used as a guide.
- Fibrillation potential (FP) in a myotome distribution may be the only abnormality.
- FP occurs in a proximal-to-distal sequence after the onset of a lesion.

*NEE of paraspinal muscles* is necessary for the proper assessment of radiculopathy.

Important considerations are:

- FP in paraspinal muscles indicates axonal lesions in the posterior primary ramus, thus indicating axonal lesions within or near the neural foramina.
- The multifidus is the deepest paraspinal muscle and is the only one considered to have monosegmental innervation.
- FP in an acute root lesion may be found 6 or 7 days after the onset of the lesion, but may not appear in limb muscle for 5–6 weeks after the lesion onset.
- FP and insertional positive sharp waves (PSW) occur in lumbar paraspinal muscles in 14.5 to 48 % of individuals.
- Paraspinal FP may be found in radiculopathies, amyotrophic lateral sclerosis (ALS), focal trauma, metastasis, and myopathies, particularly in diabetes mellitus.
- NEE may be suboptimal because of failure to achieve satisfactory muscle relaxation.
- With NEE, the extensive overlap of root innervation makes it difficult to determine the specific root involvement.
- NEE is of questionable value in the post-lumbar surgery period.

# Motor and Sensory Nerve Conduction Studies

Motor and sensory nerve conduction studies (NCS) are a helpful adjunct in the evaluation of a radiculopathy.

- The evaluation might include up to three motor NCS, to check the same nerve in the contralateral limb and another nerve in the ipsilateral limb, and two sensory NCS.
- Low compound muscle action potential (CMAP) may be due to axonal degeneration or conduction block.
- Low CMAP occurs commonly with neuropathies, but may also be present in radiculopathies when the lesion is severe or when multiple roots are involved.

- With a single-root lesion, NCS may be normal.
- In most root dysfunction, the level of injury is proximal to the dorsal root ganglion (DRG).
- In the distribution of sensory complaints or clinical sensory loss with lesions that are proximal to the DRG, normal studies of the sensory nerves suggest radiculopathy. Abnormalities of sensory nerves indicate a more distal site of injury.
- Radiculopathy and focal neuropathy may coexist in the same limb.

## Motor Evoked Responses

The responses that can be recorded from muscle after stimulation of the central nervous system or proximal nerve roots are called motor evoked potentials (MEPs). Electrical or magnetic stimulation can be used for this purpose. These studies are performed bilaterally, recording CMAPs and latencies.

## H-Reflexes and F-Waves in Evaluation of Radiculopathy

- H-reflexes can help in the evaluation of a suspected radiculopathy, while F-waves are not very helpful for this purpose.
- While abnormal H-reflexes or F-waves alone cannot establish the diagnosis of a radiculopathy, they can add to the certainty of the EDX information suggesting a diagnosis of root dysfunction.
- The H-reflex that is most commonly used and reliable is that assessing the S1 fibers in the leg.
- Side-to-side latency differences of greater than 1.0–1.8 ms, or latencies that exceed those predicted by a nomogram are used.
- Side-to-side amplitude difference of more than 50 % is also used.
- Specialized segmental H-reflex studies performed on the upper limb (flexor carpi radialis [FCR] or extensor carpi

radialis [ECR] for C6/7, biceps for C5/6, and abductor pollicis brevis [APB] for C8/T1) can be used.

- The H-reflex evaluates the function of sensory root fibers, including the segment proximal to the DRG. This evaluation is pertinent, as sensory complaints are more common than motor complaints in root lesions.

# H-Reflex in S1 Radiculopathy

Latency is the most sensitive indicator in diagnosing S1 radiculopathy. A difference in latency of one or more milliseconds in duration strongly supports the diagnosis of S1 radiculopathy if the history, physical examination, and electromyography (EMG) are compatible.

**Note:** In a prospective study, the diagnostic utility of upper limb segmental reflexes in patients with suspected CR was established. The four clinical criteria for CR were: abnormal history, motor examination, sensory examination, and changes in deep tendon reflexes. Abnormal needle EMG was found in 90 % of the subjects who had three clinical signs, and this result was found in 59 % of those with two signs, and in only 10 % of those with only one sign. H-reflexes demonstrated a sensitivity of 72 % and specificity of 85 % for the detection of CR. The specialized segmental H-reflex studies of the upper limb (FCR or ECR for C6/7, biceps for C5/6, and APB for C8/T1) were as sensitive and specific as magnetic resonance imaging.

# Motor Evoked Responses

The responses that can be recorded from muscle after stimulation of the central nervous system or proximal nerve roots are called motor evoked potentials (MEPs). Electrical or magnetic stimulation can be used for this purpose. These studies are performed bilaterally, recording CMAPs and latencies.

# Somatosensory Evoked Potentials

The use of somatosensory evoked potentials is a less common method of evaluating radiculopathies. These studies can be done by nerve trunk stimulation, cutaneous nerve stimulation, or dermatomal stimulation. The responses are recorded with surface or needle electrodes over the scalp, over the spine, and in peripheral nerve fibers in the limb under study.

# Practice Parameters for Electromyographic Needle Evaluation of Patients with Suspected Cervical Radiculopathy

The listed recommendations below relate to EDX studies when laboratory confirmation of cervical radiculopathy is needed.

1. Guideline: Needle EMG examination is performed and interpreted by a specially trained physician; the examination shows at least one muscle innervated by spinal roots C5, C6, C7, C8, and T1 in a symptomatic limb. Cervical paraspinal muscles at one or more levels, as appropriate to the clinical presentation, should be examined (except in patients with a prior cervical laminectomy, in whom a posterior approach is used). If a specific root is suspected clinically, or if an abnormality is seen on the initial needle EMG examination, additional studies are as follows:

   (a) Examination of one or two more muscles innervated by
   .    the suspected root, but by a different peripheral nerve
   (b) Demonstration of the presence of normal muscles above and below the involved root

2. Guideline: At least one motor and one sensory NCS must be performed in the clinically involved limb to check whether concomitant polyneuropathy or nerve entrapment exists. Motor and sensory NCS of the median and

ulnar nerve should be performed if symptoms and signs suggest carpal tunnel syndrome (CTS) or ulnar neuropathy. If one or more NCS are abnormal or if clinical features suggesting polyneuropathy are present, further evaluation may include NCS of other nerves in the ipsilateral or contralateral limbs to define the cause of the abnormalities.

3. Option: If needle EMG examination is abnormal, needle EMG of one or more contralateral muscles may be required to exclude bilateral radiculopathy. It may also be necessary to differentiate between radiculopathy and polyneuropathy, motor neuron disease, spinal cord lesion, and other neuromuscular disorders.

4. Option: Perform median, ulnar, or both F-wave studies in suspected C8 or T1 radiculopathy. Compare with the contralateral side if necessary.

5. Option: Perform cervical nerve root stimulation to help in identifying radiculopathy.

6. Option: Perform H-reflex study of the FCR to assist in identifying the pathology of the C6 and the C7 nerve roots.

# Lumbosacral Radiculopathy: Evidence-Based Review of EDX Testing

An evidence-based review of EDX testing was performed to determine its utility in diagnosis and prognosis in patients with suspected lumbosacral radiculopathy. In patients who might have lumbosacral radiculopathy, EDX studies with peripheral limb and paraspinal muscle EMG probably aid in its diagnosis (class II evidence, level B recommendation). The H-reflex in S1 radiculopathy showed class II and III evidence (level C recommendation), and a low sensitivity of both peroneal and posterior tibial F-waves (class II and III evidence, level C recommendation) was suggested.

However, there was inadequate evidence to reach a conclusion on the utility of dermatomal/segmental somatosensory evoked potential (SEP) of the L5 or S1 dermatomes, the utility of PM with needle EMG in sacral radiculopathy, and the utility of MEB with root stimulation in making an independent diagnosis of lumbosacral radiculopathy.

# Further Reading

Wilbourn AJ, Aminoff MJ. The electrodiagnostic examination in patients with radiculopathies. Minimonograph #32: AAEM. Muscle Nerve. 1998;21(12):1612–31.

Scho MA, Ferrante KH, Levin KH, et al. Utility of electrodiagnostic testing in evaluating patients with lumbosacral radiculopathy: an evidence-based review. Muscle Nerve. 2010;42:276–82.

Miller TA, Pardo R, Yaworski R. Clinical utility of reflex studies in assessing cervical radiculopathy. Muscle Nerve. 1999;22:1075–9.

Practice parameter for needle electromyographic evaluation of patients with suspected cervical radiculopathy: Summary Statement. AAEM and AAPM&R. Muscle Nerve. 1999;22: (Suppl 8):S209–11.

Ri B, Johnson EW. Standardization of H reflex and diagnostic use in S1 radiculopathy. Arch Phys Med Rehabil. 1974;55(4):161–6.

# Chapter 12
## Plexopathies

Plexus lesions often present a clinical and electrodiagnostic challenge. Electrodiagnostic techniques can localize a lesion to a plexus and can also provide evidence for peripheral nerve or root lesions that might produce similar symptoms. Electrodiagnostic studies can define the pathophysiology and severity of a plexopathy and can differentiate root evulsion or plexopathy related to compressions (such as from a tumor or hematoma) from plexopathy as an effect of radiation therapy. Follow-up studies show the course of the plexus injury, aiding in management and in determining prognosis.

The differential diagnoses of plexopathies are mononeuropathies and radiculopathy. It is often necessary to study all the main sensory and motor nerves that can be studied in an extremity. Extensive needle electromyography (EMG) is also required. Long latency reflexes may also provide useful information. Contralateral studies are also necessary, especially if the symptoms are bilateral or when the contralateral limb can serve as a control.

Brachial plexopathy causes weakness, sensory loss, and the loss of tendon reflexes in body regions innervated by nerves in the C5-T1 segmental distribution. The clinical diagnosis is confirmed through electrodiagnostic studies (i.e., EMG).

Lumbar plexopathy causes weakness, sensory loss, and reflex changes in regions innervated by nerves in the spinal L1-L4 segmental distribution. This results in weakness and sensory loss in the obturator- and femoral-innervated territories.

A.Q. Rana et al., *Neurophysiology in Clinical Practice*,
In Clinical Practice, DOI 10.1007/978-3-319-39342-1_12,
© Springer International Publishing Switzerland 2017

Sacral plexopathy produces the same abnormalities as lumbar plexopathy, in segments L5-S3, resulting in weakness and sensory loss in the gluteal (motor only), peroneal, and tibial nerve territories.

# Brachial Plexus Anatomy

The brachial plexus is a somatic plexus formed by the ventral rami of the lower four cervical nerves (C5–C8) plus the first thoracic nerve (T1). It innervates all of the muscles of the upper limb, with the exception of the trapezius and the levator scapula. The plexus supplies all of the cutaneous innervation of the upper limb, with the exception of the area of the axilla that is supplied by the intercostobrachial nerve. This area is just above the point of the shoulder (supplied by supraclavicular nerves) and the dorsal scapular area (supplied by cutaneous branches of the dorsal rami).

The brachial plexus communicates with the sympathetic trunk through the gray rami that join all the roots of the plexus. These rami are derived from the middle and inferior cervical sympathetic ganglia and the first thoracic sympathetic ganglion. There are two variations of brachial plexus, outlined below.

(a) *Prefixed Brachial Plexus* — happens when the C4 ventral ramus contributes to the brachial plexus. Contributions to the plexus usually come from C4–C8.
(b) *Postfixed Brachial Plexus* — occurs when the T2 ventral ramus contributes to the brachial plexus. Contributions to the plexus usually come from C6–T2.

## Etiology of Brachial Plexopathy

Brachial plexopathy is a constellation of symptoms that consist of neurogenic pain, and associated weakness that radiates from the shoulder into the shoulder girdle region and upper extremity. Brachial plexopathy has many causes; the more common ones include:

1. Compression of the plexus by ribs or muscles (thoracic outlet syndrome)
2. Invasion of the plexus by tumor (Pancoast tumor syndrome)
3. Direct trauma to the plexus (stretch injuries and avulsions, surgery)
4. Inflammatory causes (Parsonage-Turner syndrome, Herpes zoster infection)
5. Postradiation plexopathy

The clinical course of brachial plexopathy is variable, depending on the level and extent of root, trunk, and cord involvement.

Brachial plexopathy causes weakness, sensory loss, and loss of tendon reflexes in the body regions that are innervated by nerves in the C5-T1 segmental distribution. Electrodiagnostic studies (i.e., EMG) are done to confirm the clinical diagnosis.

**Electrodiagnostic Studies**

Sensory nerve action potential (SNAP) decreased
Compound muscle action potential (CMAP) decreased
EMG fibrillations (Fibs), positive sharp waves (PSW) in innervated muscles distally
Sparing of paraspinals (posterior ramus)

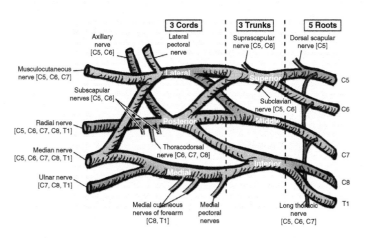

Fig. 12.1  Brachial plexus

# Lumbar Plexus

## Anatomy

The lumbar plexus arises from the psoas muscle from the anterior ramus of L1 to L4. The iliohypogastric, ilioinguinal, femoral, and lateral femoral cutaneous nerves arise from the lateral aspect of the psoas. The obturator (anterior division) nerve arises from the medial aspect of the psoas and the genitofemoral nerve arises from the posterior aspect of the psoas muscle.

## Etiology of Lumbar Plexopathy

Lumbar plexopathy can be caused by tumors, hematoma, trauma, surgical injury, diabetes mellitus, infection, vasculitis, and paraneoplastic syndromes.

The clinical course is variable, depending on the level and extent of root and trunk involvement.

Lumbar plexopathy produces sensory loss, weakness, and reflex changes in the distribution of spinal segments L1-L4. This results in weakness and sensory loss in the obturator- and femoral-innervated territories. Sacral plexopathy causes the same abnormalities as lumbar plexopathy, in segments L5-S3, causing weakness and sensory loss in the gluteal, peroneal, and tibial nerve distribution.

### Electrodiagnostic Studies

SNAP decreased
CMAP decreased
EMG Fibs, PSW in innervated muscles distally
Sparing of paraspinals (dorsal)

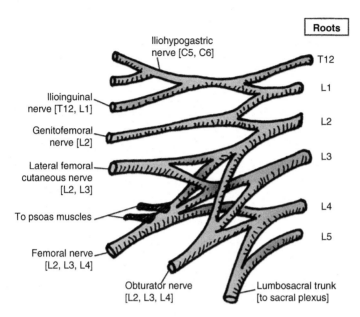

FIG. 12.2  Lumbar plexus

# Chapter 13
## Median Neuropathy

## Median Nerve Anatomy

The median nerve is formed by the combination of the medial (C8, T1) and the lateral cords (C6, C7) of the brachial plexus. The median sensory fibers innervate the thenar eminence; the thumb; and the index, middle, and lateral half of the ring finger. The motor fibers supply the proximal median forearm muscles and the muscles of the thenar eminence (opponens pollicis, abductor pollicis brevis, flexor pollicis brevis, and first and second lumbricals). The forearm muscles supplied by the median nerve are the pronator teres, flexor carpi radialis, palmaris longus, flexor pollicis longus, flexor digitorum sublimis, flexor digitorum profundus, and pronator quadratus (through the anterior interosseous nerve) see Fig. 13.1.

In the upper arm, the median nerve descends medial to the humerus and anterior to the medial epicondyle. However at the distal third of humeral shaft the median nerve is encased by the ligament of Struthers, which extends between the spur just superior to the medial epicondyle and the medial humeral epicondyle. In the antecubital fossa, the median nerve travels along and lateral to the brachial artery. As this nerve enters the forearm, it is inferior to a thick

A.Q. Rana et al., *Neurophysiology in Clinical Practice,*
In Clinical Practice, DOI 10.1007/978-3-319-39342-1_13,
© Springer International Publishing Switzerland 2017

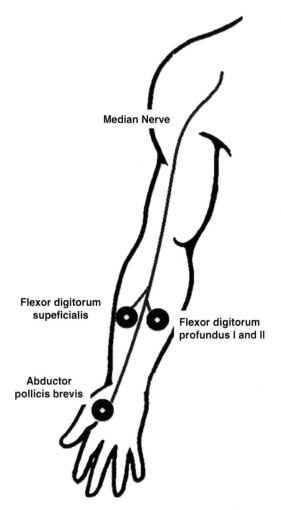

FIGURE 13.1 Median nerve neuropathy

fibrous band (lacertus fibrosus) that extends from the medial aspect of the biceps tendon to the proximal forearm flexor musculature. The nerve then runs between the two heads of the pronator teres muscle to provide innervation to that muscle. The anterior interosseous nerve arises from the median nerve approximately 5–8 cm distal to the medial epicondyle. As the median nerve runs distally, it traverses deep to the flexor digitorum sublimis. It innervates the flexor pollicis longus, the medial head of the flexor digitorum profundus to the index and the middle fingers, and the pronator quadratus muscle in the distal forearm. This nerve is a pure motor nerve. However, some deep sensory fibers are carried through the anterior interosseous nerve to supply the wrist joint.

Before the median nerve enters the carpal tunnel, it gives off a palmar cutaneous sensory branch to supply sensation over the thenar eminence. The nerve then enters the carpal tunnel along the nine flexor tendons. The sides and the floor of the tunnel are formed by the carpal bones, and the thick transverse carpal ligament forms the roof.

In the palm, the median nerve divides into motor and sensory divisions. The motor division supplies the first and second lumbricals, and the recurrent thenar motor branch supplies the abductor pollicis brevis, the opponens pollicis, and the superficial head of the flexor pollicis brevis. The sensory fibers supply the medial thumb, the entire index finger, the entire middle finger, and the lateral half of the ring finger through digital lateral and medial branches. The clinical symptoms and signs of median nerve entrapment syndromes depend on the segment of the nerve that is entrapped.

# Carpal Tunnel Syndrome

Carpal tunnel syndrome (CTS) is considered to be the most common example of median nerve entrapment, the entrapment occurring within the carpal tunnel at the wrist. CTS involves all the sensory branches to the median-innervated digits, as well as the motor branch to the first two lumbricals. CTS also involves the recurrent thenar motor branch that supplies the abductor pollicis brevis and the flexor pollicis brevis muscles. The diagnosis of this syndrome is based on history, clinical findings, and electrophysiological studies.

In a multivariate analysis, the best combination of clinical tests was found to be the tourniquet test, the carpal compression test, and Phalen's test, but the difference between these and the hand elevation test alone was negligible. Hence, the hand elevation test may be used alone, and it is superior to the questionnaires and other physical signs used in the clinical diagnosis of CTS.

The pathophysiology of CTS in the early stages is demyelination, accompanied by axonal loss in the advanced stages. The reduction in the number of axons can be estimated by the motor unit number estimation (MUNE) method.

# Conditions Associated with Carpal Tunnel Syndrome

1. Trauma with wrist fracture or wrist hemorrhage
2. Occupational and repetitive stress
3. Endocrine disorders such as hypothyroidism, acromegaly, and diabetes
4. Compressive neuropathy arising from the presence of benign tumors such as a ganglion cyst, lipoma, Schwannoma, hemangioma, or neurofibroma
5. Infectious/inflammatory conditions such as rheumatoid arthritis, septic arthritis, sarcoidosis, histoplasmosis, Lyme disease, and tuberculosis

6. Congenital conditions such as a persistent median artery, congenital small carpal tunnel, and anomalous muscles of the forearm
7. Other conditions such as spasticity causing persistent wrist flexion; hemodialysis; amyloidosis; and pregnancy; or any condition that increases fluid retention

# Electrodiagnostic Testing Methods Used Traditionally

- Stimulating the median nerve at the wrist, at the elbow just medial to the biceps tendon, in the axilla, or at Erb's point in the neck, while recording from the abductor pollicis brevis
- Recording sensory nerve action potentials from the digital nerves over the thumb, index, middle, and ring fingers in an orthodromic or an antidromic fashion
- Performing median motor studies in both extremities to compare the extent of the axonal loss
- Sensory testing for side-to-side comparisons of the median nerve, as well as ipsilateral radial or ulnar studies
- Electromyography (EMG) needle testing to include multiple median-innervated muscles, as well as other C5-T1 muscles, to rule out plexopathy or cervical radiculopathy and to ascertain the level of injury

# Diagnostic Parameters Used Traditionally

- Prolonged median terminal motor latency.
- Prolonged median terminal motor latency compared with the ipsilateral ulnar latency.
- Prolonged median terminal motor latency compared with the contralateral median latency.
- Reduced compound muscle action potential (CMAP) as recorded from the thenar eminence (conduction block or axonal loss).

- Reduced conduction velocity in the median-innervated digits.
- Relatively reduced median sensory conduction velocities compared with the ipsilateral ulnar latencies.
- The terminal latency index (TLI) is a derived measured of the distal motor conduction as a ratio of the measured distal distance to the calculated distal distance (conduction velocity × distal latency). It is calculate by the following formula

  TLI = terminal distance (mm)/conduction velocity
        (m/second) × distal latency (millisecond)

- Presence of fibrillations/fasciculations and other evidence of denervation in needle EMG testing of the median-innervated muscle.
- If two or more of the above studies are abnormal, there is a high likelihood of CTS.

# Recommendations for Electrodiagnostic Studies to Confirm a Clinical Diagnosis of CTS

1. Perform median sensory nerve conduction studies (NCS) across the wrist with a conduction distance of 13–14 cm (median sensory nerve conduction between wrist and digit).
2. If the results of the median sensory NCS are normal compared with the result of sensory NCS of one other adjacent sensory nerve in the symptomatic limb, this confirms CTS (sensitivity 0.65, specificity 0.98).
3. If the first median sensory NCS across the wrist is found to have a conduction distance of more than 8 cm and the result is normal, one of the three following additional studies is recommended:

   (a) Compare median sensory and mixed nerve conduction across the wrist over a short (7–8 cm) conduction distance (median sensory and mixed nerve conduction between wrist and palm) with ulnar sensory nerve conduction across the wrist over the same distance (sensitivity 0.74, specificity 0.97 for diagnosis of CTS).

(b) Compare median sensory conduction across the wrist with radial or ulnar sensory conduction across the wrist in the same limb (comparison of median and ulnar sensory conductions between the wrist and ring finger/comparison of median and radial sensory conduction between wrist and thumb; sensitivity 0.85, specificity 0.97 for diagnosis of CTS).

(c) Compare median sensory or mixed nerve conduction through the carpal tunnel with sensory or mixed NCS of the proximal (forearm) or distal.

4. Perform motor conduction study of the median nerve by recording from the thenar muscle (median motor nerve distal latency) and any other nerve in the symptomatic limb to include measurement of distal latency (sensitivity 0.63, specificity 0.98 for diagnosis of CTS).

5. Perform supplementary NCS, as listed below:

- Compare median motor nerve distal latency (second lumbrical) with that of the ulnar motor nerve (second interossei; sensitivity 0.56, specificity 0.98 for diagnosis of CTS), determine the median motor terminal latency index (sensitivity 0.62, specificity 0.94 for diagnosis of CTS).

- Determine the median motor nerve conduction at the wrist and the palm (sensitivity 0.69, specificity 0.98 for diagnosis of CTS).

- Determine the median motor nerve CMAP wrist-to-palm amplitude ratio to investigate for conduction block, determine the median sensory nerve action potential (SNAP) wrist-to-palm amplitude ratio to find the conduction block, and perform short-segment (1-cm incremental) median sensory nerve NCS across the carpal tunnel.

- Perform needle EMG in a sample of the muscle innervated by the C5-T1 spinal roots, including the thenar muscle, which is innervated by the median nerve of the symptomatic limb.

## Suggested Grading of CTS Severity

The greatest advantage of grading the severity of CTS in electrodiagnostic reports is in guiding the management of the syndrome. Several attempts have been made to create a grading system, using latencies, amplitudes, conduction block, and denervation. The use of a grading system identifies the extent of nerve injury and allows the referring physician to utilize the electrodiagnostic report optimally in the further management of the patient. The following grading scheme, which combines different parameters, has been suggested.

## Mild

1. Prolonged distal sensory latency (DSL) and/or median mixed nerve latency
2. Normal or minimally prolonged distal motor latency (DML)
3. Amplitudes of all responses within normal ranges
4. No conduction block, or if present, mild
5. No thenar EMG abnormalities

## Moderate

1. Prolonged DSL, mixed nerve latency (MNL), and DML.
2. Amplitudes of tested responses may be diminished, typically with relatively slight changes.
3. Conduction block might be present.
4. Minor thenar EMG abnormalities might be present.

## Severe

1. Inability to obtain median SNAPs, or low amplitude and very prolonged DSL.
2. Low amplitude or unobtainable median mixed nerve response, or if this response is present very prolonged MNL.
3. Low amplitude or unobtainable median CMAP, or if present, very prolonged DML.

4. Conduction block may be present and may be pronounced, with distal amplitude reduced by more than 70 % of normal values.
5. Thenar EMG abnormalities often present.

## Precautions/Differential Diagnoses

- Consider coexistent polyneuropathy.
- Consider cervical radiculopathy.
- Consider coexistent ulnar neuropathy.
- Diagnosis of CTS is based on the demonstration of focal slowing or conduction block of the median nerve fibers across the carpal tunnel and by excluding median neuropathy at the elbow, brachial plexopathy, cervical radiculopathy, or a coexistent polyneuropathy.

## Other Conditions That Mimic Carpal Tunnel Syndrome

### *Entrapment by Ligament of Struthers*

- A syndrome characterized by pain in the volar forearm and paresthesias in the median-innervated digits.
- Exacerbated by forearm supination and elbow extension.
- The radial pulse may also be diminished, as the brachial artery also runs along the median nerve under the ligament of Struthers.

### *Pronator Syndrome*

- Pronator syndrome is more common than entrapment at the ligament of Struthers.
- Pain radiates proximally and is exacerbated by the use of the arm, such as in repeated pronation and supination.
- Symptoms are exacerbated by resisted pronation and elbow extension or with resisted flexion of the elbow with supination of the forearm.

- There might be some weakness of the flexor pollicis longus and the abductor pollicis brevis.
- The pronator teres muscle is usually spared.
- There is paresthesia in the median-innervated digits.

### Anterior Interosseous Nerve Syndrome

- The anterior interosseous nerve (AIN) is the largest branch of the median nerve and passes between the two heads of the pronator teres.
- The AIN innervates the flexor digitorum profundus to the second and third digits, and the pronator quadratus.
- Injury to the AIN results in persistent dull, aching, volar wrist pain.
- The characteristic clinical sign is the inability to make an "OK" sign. The thumb and the index finger are unable to flex at the proximal interphalangeal joints.

## Magnetic Resonance Imaging (MRI) and CTS

- The application of MRI in CTS has been limited:
- Routine electrophysiological studies are adequate and can be performed with confidence in both community and academic settings.
- MRI remains expensive, and the
- acquisition of high-quality peripheral nerve images and their expert interpretation is not widely available.
- The low specificity of MRI could complicate treatment decisions.

## Diagnostic Ultrasound and CTS

Diagnostic ultrasound with probes working at high frequencies has proven to be useful in patients with CTS. Diffuse or localized swelling of the median nerve, due to edema and flattening by compression, is the usual cause of most of the sonographic findings.

# Case Study

A 52-year-old woman is referred for tingling and numbness over multiple digits of the right hand, with symptoms being most prominent at night; she obtains relief by rubbing and shaking the hands; symptoms are exacerbated when she drives or uses a computer.

On examination, sensations are diminished in the median nerve distribution, with no splitting of the ring finger sensations. Tinel's sign is positive but Phalen's maneuver is negative over the median nerve at the wrist. There is no thenar muscle atrophy as compared with the left hand findings.

Electrodiagnostic testing showed reduced SNAP in the median distribution as compared with the ulnar-innervated fourth and fifth digits, and prolonged median terminal motor latency of 5.4 ms on the right compared with 3.8 ms on the left. CMAP, as obtained from the right thenar eminence, was 4.8 mV, compared with 6.5 mV on the left. Needle EMG of the abductor pollicis brevis showed 2+ fibrillations and poor recruitment of the motor units.

# Further Reading

Amirfeyz R, Clark D, Parsons B, et al. Clinical tests for carpal tunnel syndrome in contemporary practice. Arch Orthop Trauma Surg. 2001;131:471–4.

Bland JDP. Do nerve conduction studies predict outcome of carpal tunnel decompression? Muscle Nerve. 2001;24:935–40.

Fleckenstein JL, Wolfe GI. MRI vs EMG Which has the upper hand in carpal tunnel syndrome? Editorial. Neurology. 2002;58(11):1583–4.

Jablecki CK, Andary MT, Floeter MK. Practice parameter: electrodiagnostic studies in carpal tunnel syndrome. Neurology. 2002;58:1589–92.

Sucher BM. Grading severity of carpal tunnel syndrome in electrodiagnostic reports: why grading is recommended. Muscle Nerve. 2013;48:331–3.

# Chapter 14
# Ulnar Neuropathy

## Ulnar Nerve Anatomy

The ulnar nerve is derived from the C8 and T1 roots with a probable minor component from C7. The ulnar fibers traverse through the lower trunk and the medial cord of the brachial plexus. The terminal extension of the medial cord is the destined ulnar nerve see Fig. 14.1.

In the upper arm, the ulnar nerve descends medially without giving off any sensory or motor branches. It pierces the intramuscular septum in the mid arm and passes through the arcade of Struthers (a band of dense fibrous tissue that extends from the medial head of the triceps) and the internal brachial ligament. The ulnar nerve travels medially and distally toward the elbow. At the elbow, it enters the ulnar groove between the medial epicondyle (ME) and the olecranon process. It travels under the arch formed by the two heads of the flexor carpi ulnaris (FCU) muscle. This is the cubital tunnel, also known as the humeral-ulnar aponeurosis.

Motor branches are given off to the FCU and the medial division of the flexor digitorum profundus for the fourth and the fifth digits. In the forearm, the nerve descends through the medial forearm up to the wrist. At about 5–8 cm proximal to the wrist, the dorsal ulnar cutaneous sensory branch is given off, and this supplies the dorsal medial hand and the

A.Q. Rana et al., *Neurophysiology in Clinical Practice*,
In Clinical Practice, DOI 10.1007/978-3-319-39342-1_14,
© Springer International Publishing Switzerland 2017

dorsal surfaces of the fourth and the fifth digits. The palmar cutaneous sensory branch is given off at the level of the ulnar styloid, and it supplies sensation to the proximal medial palm.

At the medial wrist, the nerve enters Guyon's canal and supplies sensation to the palmar aspect of the fourth and fifth digits, and muscular and motor branches are supplied to the hypothenar muscles. Also, the nerve supplies the palmar and the dorsal interossei, the third and the fourth lumbricals, and two muscles in the thenar eminence, that is, the adductor pollicis and the deep head of the flexor pollicis brevis.

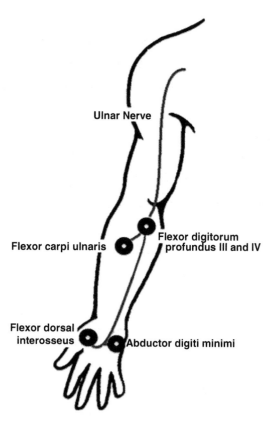

FIGURE 14.1 Ulnar nerve

# Ulnar Neuropathy at the Elbow

## *Anatomy*

SEGMENT 1. The ulnar nerve courses between the posterior compartment and the entrance to the condylar groove. It passes beneath the medial head of the triceps and the intermuscular septum (arcade of Struthers).

SEGMENT 2. Courses in the condylar groove; posterior to the ME; subluxable; compressible.

SEGMENT 3. Courses from the level of the ME and olecranon (position 0) through the fibrous arch of the FCU (fibro-osseous cubital tunnel); related to the medial ligament of the elbow. Flexion narrows the tunnel.

SEGMENT 4. Courses from between the two heads of the FCU, and passes through the aponeurosis separating the FCU from the flexor digitorum profundus and the flexor digitorum superficialis.

## *Clinical Presentation*

The ulnar nerve can be entrapped in the upper arm (arcade of Struthers), at the elbow (cubital tunnel), in the forearm, at the wrist (Guyon's canal), or in the hand.

## *Factors Predisposing to Ulnar Nerve Injury*

- Stretching and tension with elbow at 90° flexion.
- Fibrous tissue in the cubital tunnel or condylar groove.
- Positional pressure changes: increased intraneural pressure with arm elevated, elbow flexed, and wrist extended.
- Subluxation of the ulnar nerve out of the condylar groove to posterior and anterior position relative to the ME; this is related to shallow groove, short ME, blunt ME, bulging of the medial head of the triceps, or trauma to the post-condylar groove.

- Occupational factors of repetitive flexion-extension, persistent pronation, or sustained pressure, such as when driving.
- Risk in surgical patients, particularly in patients undergoing cardiac surgery.
- Predisposing factors in debilitated patients.
- Wheelchair users resting their arms on the armrests.
- Fractures and dislocations at the elbow.
- Space-occupying lesions such as ganglionic cysts, lipomas, or other tumors.
- Inflammatory disorders (rheumatoid arthritis [RA], synovitis).
- Osteoarthritis and hypertrophic bony changes.
- Double crush syndrome.
- Vascular compromise caused by surgery or vasculitis.

## Electrodiagnostic Testing

The ulnar motor component is steadied by recording from the abductor digiti minimi and stimulating the nerve at the wrist, below the elbow, at the elbow, above the elbow, at the axilla, or at Erb's point. The ulnar sensory responses are recorded from the fourth and fifth digits.

Abnormalities noted could be reduced sensory nerve action potentials (SNAPs), conduction slowing in the ulnar-innervated fourth and the fifth digital nerves, low amplitude of the ulnar compound muscle action potential (CMAP), prolonged distal latency of the ulnar nerve at the wrist, conduction block, or conduction slowing of the ulnar nerve segment at the elbow or in the forearm.

Other techniques for evaluation include inching studies across the elbow and recording the dorsal ulnar cutaneous and medial antebrachial cutaneous SNAPs, as well as performing needle electromyography (EMG) of the ulnar-innervated muscles in the hand and forearm. For the differential diagnosis, muscles supplied by the C7, C8, and the T1 roots would be studied.

Ulnar nerve conductions across the elbow segment are studied with the elbow flexed at 90°, when the true length of the ulnar nerve can be accurately measured.

## Case Study

A 45-year-old man is referred for evaluation of numbness and pain over the right hand, present for the past several months and associated with pain in the elbow. Examination shows decreased pinprick sensations in the ulnar nerve distribution and weakness of the intrinsic hand muscles. Tinel's sign is weakly positive over the ulnar nerve at the ME.

Sensory nerve conduction studies (NCS) show reduced SNAP over the fifth digit, and motor NCS show a conduction block of approximately 20 % and conduction slowing of 15 %. Needle EMG of the first dorsal interossei is unremarkable, but that of the abductor digiti minimi shows 1+ fibrillations and 2+ fasciculations, with poor recruitment of the motor units.

## Clinical Signs

- Benediction posture: This is the result of the clawing of the fourth and the fifth digits, with inability to extend at the metacarpophalangeal joints, and flexion of the distal and the proximal interphalangeal joints arising from weakness of the third and fourth lumbricals, and weakness of finger adduction due to weakness of the interossei.
- Wartenberg's sign: This is the inability to adduct the fifth digit because of weakness of the third palmar interossei.
- Froment's sign: This is the inability to pinch a paper using the finger pads of the thumb and the index finger. In an ulnar lesion, the median-innervated flexor pollicis longus and the flexor digitorum profundus contract, resulting in marked flexion of the interphalangeal joints of the thumb and the index finger.

# Ulnar Neuropathy at the Wrist

## *Anatomy*

The ulnar nerve enters Guyon's canal at the level of the distal wrist crease. The canal is formed by the pisiform bone and the hook of the hamate. Its floor is the thick transverse carpal ligament and the adjacent hamate and triquetrum bones. The roof is loosely formed.

In Guyon's canal, the nerve divides into superficial and deep branches. The motor fibers are given off as a deep palmar motor branch to three of the four hypothenar muscles. The superficial branch supplies the sensations of the palmar aspect of the fourth and fifth digits. This branch also supplies motor innervation to one of the four hypothenar muscles, the palmaris brevis. The deep palmar motor branch innervates the third and fourth lumbricals, the four dorsal and the three palmar interossei, the abductor pollicis, and the deep head of the flexor pollicis brevis.

- With a lesion of the distal deep palmar motor branch, all the ulnar-innervated muscles of the hand are affected, except for the hypothenar muscles, the motor innervation to the palmaris brevis, and the sensory fibers. With a lesion to the proximal deep palmar motor nerve, all the ulnar-innervated hand muscles are affected, including the hypothenar muscles. However, the palmaris brevis and the sensory nerves are spared.
- If the lesion occurs at the proximal Guyon's canal, it affects all branches of the ulnar nerve, including the proximal and the distal deep palmar motor and sensory nerves, as well as the motor innervation to the palmaris brevis. However, if the lesion affects only the superficial sensory branch, the motor nerves are spared.
- Clinically, there is weakness of thumb abduction, weakness of the fourth and fifth flexor digitorum profundus muscles, and weakness of index finger extension.

## Electrodiagnostic Testing

Motor NCS are performed by stimulating the ulnar nerve at the wrist and around the elbow with recording from the abductor digiti minimi. Sensory nerve conduction studies are carried out on the fourth and fifth digits. The dorsal ulnar cutaneous nerves can also be studied. The ulnar nerve can also be studied by recording from the first dorsal interossei. Needle EMG of the ulnar-innervated muscles would show findings of denervation, reinnervation, or poor recruitment.

## Case Study

A 45-year-old right-handed man complained of a right arm pain associated with tingling and numbness over the fourth and fifth digits. He also complained of vague pain in the right elbow. On examination, there was relative atrophy of the intrinsic muscles of the right hand as compared with the left, and weakness of the abductor digiti minimi. Sensory examination was normal.

Ulnar motor conduction studies showed reduced CMAP as recorded from the abductor digiti minimi, with moderately prolonged distal motor latency. In comparison, the amplitude on the left side, as obtained from the abductor digiti minimi, was approximately 40 % more than that on the right side. Needle EMG showed 1+ fibrillation in the first dorsal interossei (FDI), but not in the abductor digiti minimi.

## Electrodiagnostic Studies of Ulnar Neuropathy at the Elbow

The following recommendations are made for the electrodiagnostic evaluation of patients with suspected UNE.

## General Principles

1. Ulnar sensory and motor NCS should be performed with surface stimulation and recording. Limb temperature should be monitored and maintained within a reference range; it should be reported if outside a reference range. Corrections in conduction for temperature, if any, should be indicated in the report, although warming, cooling, and repeating the studies is preferred when possible.
2. If ulnar sensory or motor NCS are abnormal, further NCS should be carried out to exclude a diffuse process.

## Elbow Position

1. The ulnar motor NCS report should specify the elbow position used during the performance of the studies and the reference values that were employed. The technique for the study should be the same as that used to determine the reference values. The same elbow position should be employed during both stimulation and measurement.
2. The most logical elbow position for ulnar NCS is moderate flexion; 70° to 90° from the horizontal level. Moderate flexion provides the best correlation between surface skin measurement and true nerve length.
3. Studies performed have used across-elbow distances, where the elbow is in moderate flexion; these distances have been in the range of 10 cm and this correlates best with the published reference values.
4. Stimulation at a distance of more than 3 cm distal to the ME had better not be done, as the nerve is usually deep within the FCU muscle at this point, and there is a substantial risk of submaximal stimulation.

## Technique

1. When using moderate elbow flexion, a 10-cm across-elbow distance, and surface stimulation recording, the following factors are suggestive of a focal lesion involving the ulnar

nerve at the elbow: multiple internally consistent abnormalities are more convincing than isolated abnormalities, which raise the possibility of artifact or technical mishap:

(a) Absolute motor nerve conduction velocity (NCV) from above the elbow (AE) to below the elbow (BE) of less than 50 m/s.

(b) An AE-to-BE segment NCV greater than 10 m/s slower than that of the BE-to-wrist (W) segment. The literature is inadequate to come up with a recommendation regarding the percentage of slowing.

(c) A decrease in compound muscle action potential (CMAP) negative peak amplitude from BE to AE that is greater than 20 %; this suggests conduction block, or a temporal dispersion indicating a focal demyelination. A conclusion from these findings that UNE is present presumes that anomalies of innervation, such as Martin-Gruber anastomosis, are not present.

(d) A significant change in CMAP configuration at the AE site compared with that at the BE site. A conclusion from these findings that UNE is present presumes that anomalies of innervation, such as Martin-Gruber anastomosis, are not present.

(e) Nerve action potential (NAP) recording may aid in diagnosis, especially in patients with only sensory symptoms. However, NAP studies have significant pitfalls and limitations. Before relying on changes in NAP amplitude or NCV as a diagnostic criterion for UNE, the examiner should be very aware of the content of, and all the technical details in the applicable literature. Abnormalities of distal sensory or mixed NAP, especially loss of amplitude, are nonspecific and nonlocalizing features of UNE.

(f) The literature is not adequate to come up with a recommendation regarding conductions through the AE-to-W or BE-to-W segments.

2. In ulnar conduction studies with stimulation at the wrist where AE and BE recordings from the abductor digiti quinti are found to be inconclusive, the following procedures may be of benefit:

(a) NCS recorded from the FDI muscle. Because of differential fascicular involvement, fibers to the FDI may show abnormalities that are not evident when recording from the abductor digiti minimi.

(b) An imaging study, exploring for changes in the CMAP amplitude, configuration, area, or for abnormal changes in latency over precisely measured 1- or 2-cm increments from AE to BE. The most convincing abnormality involves both a change in latency and a change in either amplitude, configuration, or area; however, latency changes in isolation may be significant.

(c) With severe UNE, distal Wallerian degeneration may slow the BE-to-W segment secondarily and make localization difficult. Comparison of the AE-to-BE segment with the axilla-to-AE segment may be helpful under such conditions, but normative data is scarce.

(d) Generally, NCS of forearm flexor muscles are not helpful, but they may be employed as a last resort, with full awareness of the technical limitations and the applicable literature.

(e) Depending on the results of NCS, needle EMG may be indicated. Needle examination must always include the FDI muscle – which is the most frequent one to show abnormalities in UNE – and the forearm flexor muscles innervated by the ulnar nerve. Neither changes limited to the FDI nor sparing of the forearm muscles eliminates an elbow lesion. If the ulnar-innervated muscles are abnormal, then the examination must be extended to include non-ulnar C8/medial cord/lower trunk muscles, to eliminate brachial plexopathy, and the cervical paraspinals, to eliminate radiculopathy.

# Chapter 15
# Radial Neuropathy

## Radial Nerve Anatomy

The radial nerve derives its innervation from the C5-T1 nerve roots and all the three trunks of the brachial plexus. The posterior divisions from all the three trunks unite to form the posterior cord. This posterior cord gives off the axillary, thoracodorsal, and subscapular nerves before it forms the radial nerve.

In the proximal upper arm, the radial nerve gives off the posterior cutaneous nerve of the arm, the lower lateral cutaneous nerve of the arm, and the posterior cutaneous nerve of the forearm. The muscular branches are then given off to the three heads of the triceps and the anconeus muscle at the elbow, then the radial nerve wraps around the posterior humerus in the spiral groove and descends into the region of the elbow. Muscular branches are given off to the brachioradialis and the extensor carpi radialis (longus and brevis). Approximately 3–4 cm distal to the lateral epicondyle the radial nerve divides into a superficial and a deep branch. The superficial branch is the superficial radial sensory nerve, which travels distally into the forearm over the radial bone and supplies sensation to the lateral dorsum of the hand, as well as part of the thumb and the dorsal proximal phalanges of the index, middle, and ring fingers. This superficial radial sensory nerve terminates over the thumb see Fig. 15.1.

A.Q. Rana et al., *Neurophysiology in Clinical Practice*,
In Clinical Practice, DOI 10.1007/978-3-319-39342-1_15,
© Springer International Publishing Switzerland 2017

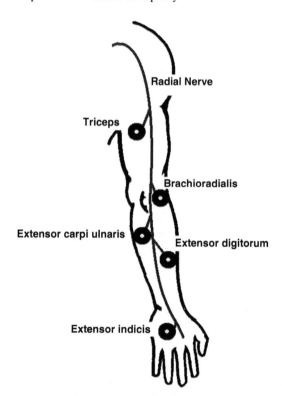

Figure 15.1 Radial nerve

The deep radial motor branch supplies the extensor carpi radialis brevis and the supinator muscles before it passes under the arcade of Frohse, which can be a site of entrapment. This arcade is formed by the proximal border of the supinator muscle. As the nerve enters the supinator, it is called the posterior interosseous nerve (PIN). This nerve then supplies all the extensors of the wrist and the digits, such as the extensor digitorum, extensor carpi ulnaris, abductor pollicis longus, extensor indicis proprius, extensor pollicis longus, and extensor pollicis brevis. While there are no cutaneous sensory branches in the PIN, it does supply sensory fibers to the interosseous membrane and the radioulnar joints.

# Wrist Drop

When the radial nerve is injured at the spiral groove a wrist drop results. This is a frequent presentation when a person wraps their arm over a chair or a bench during deep sleep or when in a state of intoxication (Saturday night palsy). A wrist drop can also result from strenuous muscular effort, fracture of the humerus, injury to the posterior cord, C7 radiculopathy, or from a vasculitic lesion. A wrist drop can also occur with the isolated involvement of the PIN that involves only the motor fibers and spares the superficial radial sensory nerve.

# Electrodiagnostic Testing

## Radial Motor Neuropathy

The radial motor nerve is studied by recording from the extensor indicis proprius with recording electrode R1 placed 4 cm proximal to the ulnar styloid and recording electrode R2 over the ulnar styloid. The radial nerve is stimulated in the forearm, at the elbow in the sulcus between the biceps and the brachioradialis, and below and above the spiral groove. With axonal loss, the compound muscle action potential (CMAP) will be reduced and motor nerve conduction velocity will be slowed, or there may be a conduction block.

Needle electromyography (EMG) of the muscles supplied by the radial nerve (triceps, anconeus, brachioradialis, and extensor carpi radialis) and the muscles supplied by the PIN (extensor digitorum, extensor indicis, and extensor carpi ulnaris) may show fibrillation potentials as evidence of axon loss. To exclude other causes of wrist drop, the posterior cord-innervated latissimus dorsi, the axillary-innervated deltoid, and other C7 myotome muscles outside the posterior cord or radial nerve (pronator teres, flexor carpi radialis) are also examined.

## Radial Sensory Neuropathy

The superficial radial nerve can be studied by orthodromic or antidromic techniques. With the orthodromic technique, stimulation is performed over the digital branches on the dorsum of the thumb, with recording 14 cm proximally along the nerve, using surface electrodes. The forearm is neutral in position and the recording electrodes are placed over the crest over the radius. A hand-held surface stimulator should be used, rather than ring electrodes, to reduce the overflow of median nerve stimulation. Alternatively, the radial nerve can be stimulated at the wrist for conduction studies of more proximal segments of the nerve. Antidromic techniques have also been described for the distal segment, with stimulation at the wrist and recording with ring electrodes at the thumb.

## Posterior Interosseous Neuropathy

Posterior interosseous neuropathy presents as a wrist drop with sparing of the brachioradialis and the extensor carpi radialis longus and brevis. Hence, the wrist can be extended partially with a radial deviation. There is no cutaneous sensory loss. Posterior interosseous neuropathy commonly occurs as an entrapment under the arcade of Frohse, but it can also be due to any mass lesion.

Motor nerve conduction study is undertaken by stimulating the nerve between the elbow and the forearm and recording from the extensor indicis proprius. The CMAP will be reduced and motor nerve conduction velocity slowed, or there may be a conduction block.

Needle EMG of the muscles supplied by the PIN (extensor digitorum, extensor indicis) may show fibrillation potentials as evidence of axon loss. To exclude other causes of wrist drop, the radial-innervated extensor carpi radialis, anconeus, and triceps and the axillary-innervated deltoid and other C7 myotome muscles outside the posterior cord or radial nerve (pronator teres, flexor carpi radialis) are also examined.

# Case Study

A 55-year-old man was referred for evaluation of a left wrist drop. He woke up one morning with a complete wrist drop without associated pain. He did have abnormal sensations over the back of his left hand between the thumb and the index finger. Physical examination showed a complete left wrist drop. Wrist and finger flexion motions were intact. Axillary nerve function, as well as that of the posterior cord, as noted by shoulder abduction, was normal. Elbow extension was normal, suggesting normal innervation to the triceps.

Radial nerve motor nerve conduction studies showed decreased CMAP when the extensor indicis proprius was stimulated over the forearm, as compared with findings on the right side. Conduction velocities were also slow when the radial nerve segment was stimulated between the spiral groove and the elbow. Needle EMG showed fibrillations in the extensor indicis proprius and the extensor digitorum. The triceps and the anconeus muscles were normal.

# Further Reading

Moore FG. Radial neuropathies in wheelchair users. Am J Phys Med Rehabil. 2009;88(12):1017–9.

Shobha N, Taly AB, Sinha S, Venkatesh T. Radial neuropathy due to occupational lead exposure: Phenotypic and electrophysiological characteristics of five patients. Ann Indian Acad Neurol. 2009;12(2):111–5.

# Chapter 16
## Lower Extremity Neuropathies

## Sciatic Neuropathy

Sciatic neuropathies typically present with weakness and sensory loss in the sciatic nerve distribution over a lower extremity. It is not unusual for the peroneal distribution to be affected much more than the tibial division, because the peroneal division has few large fascicles with relatively little intervening fibrous tissue and the tibial division has many small fascicles cushioned by a large amount of fibrous tissue [2]. Sciatic neuropathy can result from hip surgery and may also result from injections in the gluteal muscles. Piriformis syndrome is another possible etiology of sciatic neuropathy, as in about 6% of cadaver specimens the sciatic nerve passes within the piriformis muscle [1]. An important differential diagnosis is an injury to the L5, S1 roots, or a lumbar plexus lesion see Fig. 16.1.

### Diagnostic Studies

1. Motor nerve conduction studies of the peroneal and the tibial nerves are likely to show reduced compound muscle action potentials (CMAPs) compared with those obtained from the corresponding distal muscles. This helps in determining the degree of axonal loss but is not good in localizing the lesion. Similarly, late responses such as F-waves or the H-reflex can help to determine whether both the tibial

A.Q. Rana et al., *Neurophysiology in Clinical Practice*,
In Clinical Practice, DOI 10.1007/978-3-319-39342-1_16,
© Springer International Publishing Switzerland 2017

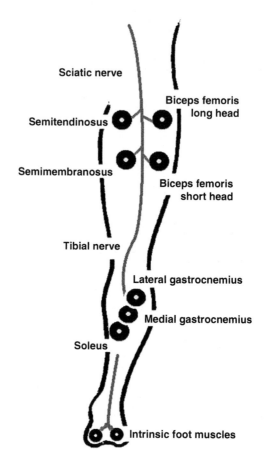

FIGURE 16.1 Sciatic nerve

and the peroneal branches are affected. Direct stimulation of the sciatic nerve at the gluteal fold is not a reliable technique for localizing sciatic neuropathy.

2. Sensory nerve action potentials (SNAPs) of the sural and superficial peroneal nerves are useful to distinguish a postganglionic lesion (example: sciatic nerve or lumbosacral plexus) from a preganglionic lesion (example: root or cauda equina).

3. Needle electromyography is useful for evaluating denervation or reinnervation after complete or very severe lesions.

Muscles distal to the lesion such as the hamstrings, where the hamstring muscles could be examined, as well as the muscles innervated by the posterior tibial nerve and the common peroneal nerve. In a sciatic lesion, the gluteal muscles are generally spared and so also are the paravertebral muscles.

## Case Study

A 58-year-old man sustained an injury to his right hip during a motor vehicle accident, with radiographs revealing an acetabular fracture treated with open reduction internal fixation (ORIF). Following the surgery, the patient woke up with a right foot drop.

Physical examination showed weakness in hip abduction, knee flexion, ankle plantar flexion, and ankle inversion, but most prominently weakness of ankle dorsiflexion. Sensations were reduced over the dorsum of the foot and the lateral aspect of the leg. Reflexes were normal at the knees but absent at the right ankle. Differential diagnoses included sciatic neuropathy, peroneal neuropathy, and L5 radiculopathy.

# Fibular Neuropathy

## Anatomy

The common fibular nerve is derived from L4, L5, S1; the fibers travel through the sciatic nerve, and at the mid-thigh level are segregated from the tibial fibers. The first branch of the common fibular nerve, while it is still bundled as the sciatic nerve, innervates to the short head of the biceps femoris. At the proximal popliteal fossa the common fibular nerve emerges as a distinct nerve. The first branch, at the level of the knee, is the lateral cutaneous nerve to the knee. The nerve then curves around the fibular neck, dividing into the superficial and deep muscular branches. The superficial muscular branch innervates the fibularis longus and fibularis brevis, and its terminal branch is the sensory superficial fibular nerve supplying the outer leg and dorsum of the foot, sparing the first web space of the foot. The deep fibular muscular branch innervates the fibularis tertius, tibialis anterior, extensor digitorum longus, extensor hallucis longus, and extensor digitorum brevis (EDB); the terminal sensory branch supplies the first web space.

## Clinical Presentation

The common presentation is that of a foot drop, with tingling and numbness of the leg and foot.

## Etiology

Fibular neuropathy is commonly caused by compression (cast, brace, crossing legs, deep squats), entrapment, or ischemia, or by trauma from a direct contusion over the fibular head, or perioperatively from knee or hip surgery.

## Electrodiagnosis

The fibular motor nerve is stimulated at the ankle, fibular head, and the lateral border of the popliteal fossa, with the CMAP recorded from the EDB. The tibialis anterior is a useful alternative muscle to record from. The amplitude of the CMAP and any temporal dispersion is recorded, along with any conduction slowing or conduction block. Comparison with the contralateral side, with a drop of >50 % in conduction velocity (CV) is considered significant. Needle electromyography examination (NEE) of the multiple muscles innervated by the superficial and the deep branches is undertaken. A muscle above the knee and outside the fibular nerve distribution is also examined for the differential diagnosis of L5 radiculopathy and sciatic neuropathy.

The superficial fibular sensory nerve study is undertaken by antidromic stimulation of the lateral calf and recording from the lateral ankle. Comparison with the asymptomatic contralateral side is generally helpful.

## Case Study

A 54-year-old man reported the sudden onset of right foot drop 3 weeks prior to presentation; the onset was noted after a weekend that he had spent laying tiles on his basement floor. He now has difficulty in walking and a tendency to fall because of the right foot drop. There is no back pain. Bowel and bladder function is normal. There is sensory loss over the top of the right foot and along the lateral aspect of the right leg.

Physical examination reveals 0/5 ankle dorsiflexion, 2/5 ankle eversion, 5/5 ankle inversion, and 5/5 ankle plantar flexion. Strength in the proximal muscles around the hip and the knee is normal. Sensations are reduced to pinprick and touch over the dorsum of the right foot, including the first web space, and over the lateral aspect of the right leg distally. Reflexes are normal.

Right fibular motor nerve conduction study showed normal CMAP of 6.5 mV when the nerve was stimulated at the ankle, 5.9 mV when stimulated at the fibular neck, and 2.3 mV when stimulated at the lateral border of the popliteal fossa. Conduction velocity (CV) over the leg segment was 46 m/s and CV across the knee was 32 m/s. NEE showed +2 fibrillations in the tibialis anterior and fibularis longus, but no fibrillation was shown in the medial gastrocnemius, tibialis posterior, and short head of the biceps femoris.

The electrophysiological studies showed evidence of an acute fibular neuropathy at the fibular neck, with demyelination and axonal loss.

# Tarsal Tunnel Syndrome

## Anatomy

The terminal portion of the tibial nerve courses posterior to the medial malleolus, and under the flexor retinaculum. It branches into the medial plantar, lateral plantar, and calcaneus branches.

## Etiology

Common causes of tarsal tunnel syndrome are direct trauma, space-occupying lesions, or biomechanical factors related to foot deformities. Often the syndrome is idiopathic.

## Electrodiagnosis

The tibial nerve is stimulated posterior to the medial malleolus and recording is done from the abductor hallucis muscle. Onset latency is prolonged when compared with the contralateral side. The CMAP is obtained from the abductor hallucis brevis. The SNAP obtained from the medial and lateral plantar nerves is diminished. NEE may reveal fibrillation potentials and positive sharp waves from intrinsic foot muscles.

# References

1. Pezina M. Contribution to the etiological explanation of the piriformis syndrome. Acta Anat. 1979;105:181.
2. Sunderland S. Nerves and nerve injuries. 2nd ed. Edinburgh: Churchill-Livingstone; 1968.

# Chapter 17
## Amyotrophic Lateral Sclerosis

## Amyotrophic Lateral Sclerosis

Electrodiagnostic tests are essential for the diagnostic workup of a patient with suspected amyotrophic lateral sclerosis (ALS). It is also important to exclude other conditions such as myopathy, neuromuscular transmission disorders, demyelinating polyneuropathy, plexopathy, and radiculopathy.

### *Lambert Criteria [2] (1969)*

First, the sensory nerve conduction studies are normal.

Second, the motor nerve conduction velocities (CVs) that are recorded from relatively unaffected muscles are normal, and when the CVs are recorded from severely affected muscles they are not less than 70 % of the average normal value.

Third, fibrillation and fasciculation potentials are present in muscles of the upper and lower extremities or in the muscles of the head and the extremities.

Fourth, motor unit potentials are reduced in number and increased in duration and amplitude.

A.Q. Rana et al., *Neurophysiology in Clinical Practice*,
In Clinical Practice, DOI 10.1007/978-3-319-39342-1_17,
© Springer International Publishing Switzerland 2017

## Revised El Escorial Criteria for ALS [1] (1994)

Needle electromyography (EMG) examination shows evidence of active and chronic denervation in at least two of the four following regions: (1) brain stem, (2) cervical, (3) thoracic, and (4) lumbosacral. For the brain stem region, needle EMG abnormalities are required in one muscle only. For the thoracic region, abnormalities can be in the paraspinal muscles below the T6 level or in the abdominal muscles. For the cervical and the lumbosacral paraspinal regions, the abnormalities must be present in more than two muscles innervated by two different nerve roots and peripheral nerves. Needle examination of the sternocleidomastoid muscle has sensitivity similar to that for examination of the tongue in patients with bulbar symptoms [3].

## Airlie House Criteria (2008)

### Principles

To make the diagnosis of ALS: First there should be evidence of lower motor neuron (LMN) degeneration shown by clinical, electrophysiological, or neuropathological examination. Second, there should be evidence of upper motor neuron (UMN) degeneration shown by clinical examination, and progressive spread of symptoms or signs within a region or to other regions, as determined by history, physical examination, or electrophysiological tests.

Furthermore, there should not be any electrophysiological or pathological evidence of other disease processes that might explain the signs of UMN or LMN degeneration. Also, there is no neuroimaging evidence of other disease processes that might explain the observed clinical and electrophysiological signs.

## Diagnostic Categories

*Clinically definite ALS* is defined by clinical or electrophysiological evidence of the presence of LMN and UMN signs in the bulbar region and in at least two spinal regions, or the presence of LMN and UMN signs in three spinal regions.

*Clinically probable ALS* is defined by clinical or electrophysiological evidence of LMN and UMN signs in at least two regions, with some UMN signs necessarily above the LMN signs.

*Clinically possible ALS* is defined when clinical or electrophysiological signs of UMN and LMN dysfunction are found in only one region; or UMN signs are found alone in two or more regions; or LMN signs are found above the UMN signs. Neuroimaging and clinical laboratory studies will have been performed and other diagnoses must have been excluded.

## *Criteria for Detection of Neurogenic Changes by Needle EMG in Diagnosis of ALS*

First, for the evaluation of lower motor neuron disease in ALS in any given body region, clinical and electrophysiological abnormalities have equal diagnostic significance. Second, EMG features of chronic neurogenic changes must be found; such as: (1) motor unit potentials (MUPs) of increased amplitude and duration are usually present with an increased number of phases, as assessed by qualitative or quantitative studies; (2) there is reduced motor unit recruitment, defined by the rapid firing of a reduced number of motor units. In limbs affected by clinical features of significant upper motor neuron abnormalities, rapid firing may not be achieved; and (3) unstable and complex MUPs are present in most cases of ALS when a narrow band pass filter (500 Hz to 5 KHz) is used). Third, in ALS, fibrillation and positive waves are usually recorded in strong, non-wasted muscles. Finally, the presence of chronic neurogenic changes on needle EMG in ALS, such as fasciculation potentials, preferably of complex morphology, is equivalent in clinical significance to fibrillations and positive sharp waves.

## Use of Nerve Conduction Studies in ALS: Exclusion of Other Disorders

The following findings are compatible with ALS:

1. Normal sensory nerve action potential (SNAP) amplitude and normal sensory CVs in the absence of concomitant entrapment or other neuropathies. Mildly reduced SNAP amplitudes and CVs in the presence of neuropathy of identifiable etiology are acceptable.
2. Motor CV > 75 % of the lower limit of normal, and minimum F-wave latency < 130 % of the upper limit of normal.
3. Distal compound muscle action potential (CMAP) latency and duration < 150 % of normal.
4. The absence of conduction block (CB) and of pathological temporal dispersion, as defined by baseline-negative CMAP, and a reduction of proximal-versus-distal stimulation of more than 50 % when distal baseline-negative peak CMAP amplitude is large enough to allow such assessment (usually more than 1 mV). A proximal negative peak CMAP duration and < 30 % of the distal value suggest CB.

## Case Study

A 70-year-old man presented to the clinic with a history of progressive right lower limb weakness. He had been dragging his left foot for the past 8 months and had had magnetic resonance imaging (MRI) of the lumbar spine; this showed mild degenerative disc disease and he had been treated conservatively. The weakness had become worse and had spread to the other limb as well, and he was wheelchair-bound on presentation. He had also noticed that his right hand was becoming weak and he could no longer push his wheelchair. And he was having difficulty in swallowing. On examination he had brisk reflexes throughout, bilateral ankle and patellar clonus, bilateral upgoing toes, and Hoffman's sign. Strength examination revealed Medical Research Council muscle strength grade 0/5 in the lower limbs and 3/5 in the upper limbs. Sensory examination was normal. EMG (shown below) revealed denervation in the cranial, cervical, thoracic, and lumbar regions, consistent with a history of motor neuron disease (note; ALS is a clinical diagnosis supported by EMG) see Table 17.1 for typical EMG findings in ALS.

## Bulbospinal Muscular Atrophy

Bulbospinal muscular atrophy is also known as Kennedy's disease. It is a rare recessive disorder that is linked to the X chromosome, and it is commonly transmitted through females. However, Kennedy's disease is only seen in adult males, and the onset of this disease is typically later in life. Symptoms of bulbospinal muscular atrophy include weakness and wasting of the facial, bulbar, and extremity muscles. There are also sensory and endocrinological disturbances, such as gynecomastia and reduced fertility. Additionally, features such as abnormal conduction through motor and sensory nerves, neuropathic (or in rare cases myopathic) alterations in biopsies of muscle cells, elevated levels of testosterone and other sex hormones, and the development of hyper-creatine kinase (CK)-emia can be seen.

TABLE 17.1 Typical EMG findings in a patient with ALS

| | Spontaneous | | | | | Motor unit action potential | | | Rec |
|---|---|---|---|---|---|---|---|---|---|
| | Insertional activity | Fibrillations | Positive wave | Fasciculation | Myotonia | Amplitude | Polyphasia | Duration | Recruitment |
| Left Trapezius, middle | Normal | +1 | +1 | +1 | | N | N | N | N |
| Left Paraspinal, thoracic | Normal | +1 | +1 | +1 | | | | | |
| Tongue | Normal | 0 | 0 | +1 | | N | N | N | N |
| Left Extn. digitorum com | Normal | +1 | +1 | +2 | | N | N | Long | N |
| Left Deltoid, middle | Normal | +2 | 0 | 0 | | N | N | Long | N |
| Complex fasciculation | | | | | | | | | |
| Left Abduc. pol. brevis | Normal | +2 | +2 | 0 | | N | N | Long | Red |
| Left First dorsal inter. | Normal | +2 | +2 | +2 | | N | N | Long | Red |
| Left Biceps brachii | Normal | +1 | 0 | 0 | | N | N | N | N |
| Left Semitendinosus | Normal | +1 | +1 | +1 | | N | N | N | N |
| Left Vastus medialis | Normal | +1 | +1 | +1 | | N | N | Long | N |
| Left Tibialis anterior | Normal | +2 | +2 | +1 | | N | N | Long | Red |

# References

1. Brooks BR. World Federation of Neurology Subcommittee on motor neuron disease. El Escorial WFN criteria for the diagnosis of amyotrophic lateral sclerosis. J Neurol Sci. 1994;124:965–1085.
2. Lambert E. Electromyography in amyotrophic lateral sclerosis. In: Norris FH, Kurland LT, editors. Motor neuron disease: Research in amyotrophic lateral sclerosis and related disorders. New York: Grune and Stratton; 1969. p. 135–53.
3. Li J, Petajan J, Smith G, Bromberg M. Electromyography of sternocleidomastoid muscle in ALS: a prospective study. Muscle Nerve. 2002;25(5):725–8.

# Further Reading

De Carvalho M, et al. Electrodiagnostic criteria for the diagnosis of ALS. Clin Neurophysiol. 2008;119:497–503.

World Federation of Neurology Research Group on Neuromuscular Diseases Subcommittee on Motor Neuron Disease. Airlie House guidelines. Therapeutic trials in amyotrophic lateral sclerosis. Airlie House "Therapeutic Trials in ALS" Workshop Contributors. J Neurol Sci. 1995;129 Suppl:1–10.

# Chapter 18
## Autonomic Neuropathy

The autonomic nervous system (ANS) maintains cardiovascular, gastrointestinal, genitourinary, exocrine, pupillary, and thermoregulatory function. Testing ANS function is an important area of clinical neurophysiology. Conventional nerve conduction studies are not helpful because they do not evaluate the small fibers.

## Parasympathetic Cardiovagal Function

Tests of parasympathetic cardiovagal function include the analysis of heart rate response to standing (the 30:15 ratio), heart rate variation with deep breathing, and the Valsalva ratio. Breathing tests measure the heart rate and blood pressure response to breathing exercises such as the Valsalva maneuver.

## Sympathetic Adrenergic Vascular Function

Tests of sympathetic adrenergic vascular function include blood pressure analysis while standing, the Valsalva maneuver, sustained handgrip, and cold water immersion. The tilt-table test monitors the blood pressure and heart rate response to changes in posture and position, simulating what occurs

A.Q. Rana et al., *Neurophysiology in Clinical Practice*,
In Clinical Practice, DOI 10.1007/978-3-319-39342-1_18,
© Springer International Publishing Switzerland 2017

when the patient stands up after lying down. The patient lies flat on a table, which is then tilted to raise the upper part of the body. Normally, the body compensates for the drop in blood pressure that occurs when a person stands up by narrowing the blood vessels and increasing the heart rate. This response may be slowed or abnormal if the person has autonomic neuropathy. A simpler way to test for postural changes in blood pressure involves standing for a minute, then squatting for a minute, and then standing again. Blood pressure and heart rate are monitored throughout this test. Breathing tests measure how the heart rate and blood pressure respond to breathing exercises such as the Valsalva maneuver.

## Sympathetic Cholinergic Sudomotor Function

Tests of sympathetic cholinergic sudomotor function include the sympathetic skin response (SSR), quantitative sudomotor axon reflex test, sweat box testing, and quantification of sweat imprints. The quantitative sudomotor axon reflex test (QSART) evaluates the nerves that regulate the sweat glands' response to stimulation. A small electrical current is passed through capsules, placed on the patient's forearm, foot, and leg, while a computer analyzes how the nerves and sweat glands react. The patient may feel warmth or a tingling sensation during the test.

## Thermoregulatory Sweat Test

During this test, the patient is coated with a powder that changes color with sweating. The patient enters a chamber with slowly increasing temperature, which will eventually make the patient perspire. Digital photos document the results. The sweat pattern may help confirm a diagnosis of autonomic neuropathy or other causes of decreased or increased sweating.

# Pupil Function

Pupil function is tested pharmacologically and with pupillographic techniques.

# Sympathetic Skin Response

The SSR is a potential generated in skin sweat glands. This response originates from activation of the reflex arch by different kinds of stimuli. The response typically produces biphasic or triphasic slow wave activity with relatively stable latency and variable amplitude. In healthy subjects younger than 60 years of age the response is always present in all extremities. The SSR is most frequently used in diagnosing the functional impairment of non-myelinated postganglionic sudomotor sympathetic fibers in peripheral neuropathies, but it is a poor man's test, with inadequate sensitivity and specificity for autonomic dysfunction, and there is no close correlation between the presence or absence of the SSR and the severity of autonomic dysfunction. The reader is recommended to read the article by Claus et al. to gain further understanding of the methodology and interpretation of autonomic studies (http://www.clinph-journal.com/pb/assets/raw/Health%20Advance/journals/clinph/Chapter7.1.pdf).

# Gastrointestinal and Genitourinary Function

Tests of gastrointestinal and genitourinary function do not satisfactorily isolate autonomic regulation from the other functions of the gastrointestinal and genitourinary tracts. The available tests are typically administered in a battery, which improves reliability and sensitivity. Gastric-emptying tests are the most common tests used to check for slowed movement of food through the system, delayed emptying of the stomach, and other abnormalities. These tests are usually performed by a gastroenterologist.

# Autonomic Neuropathies

Autonomic neuropathies can be classified as either hereditary or acquired. Acquired autonomic neuropathies may be
subdivided into primary and secondary types.

1. Hereditary autonomic neuropathies:

    (a) Familial amyloid polyneuropathy
    (b) Hereditary sensory autonomic neuropathy (HSAN
        type 1 to type 5)
    (c) Fabry disease
    (d) Acute intermittent porphyria

2. Acquired autonomic neuropathies:

    (a) Primary acquired autonomic neuropathies:

        1. Pandysautonomia
        2. Idiopathic distal small fiber neuropathy
        3. Holmes-Adie syndrome and Ross syndrome
        4. Chronic idiopathic anhydrosis
        5. Amyloid neuropathy
        6. Postural orthostatic tachycardia syndrome

    (b) Secondary acquired autonomic neuropathies:

        1. Diabetes mellitus.
        2. Uremic neuropathy.
        3. Hepatic disease-related neuropathy.
        4. Vitamin deficiency and nutrition-related neuropa-
           thy (example: B12 neuropathy).
        5. Toxin- and drug-induced autonomic neuropathy
           (example: neuropathy caused by the chemothera-
           peutic agents vincristine, cisplatin, and paclitaxel).
        6. Alcohol-associated autonomic neuropathy.
        7. Infectious diseases may be associated with auto-
           nomic neuropathy (examples are: Lyme disease,
           HIV infection, Chagas disease, botulism, diphtheria,
           and leprosy).

8. Autoimmune conditions may be associated with autonomic neuropathy (examples are: celiac disease; Sjögren's syndrome, rheumatoid arthritis, and connective tissue diseases; Guillain-Barré syndrome; Lambert-Eaton myasthenic syndrome; paraneoplastic autonomic neuropathy; and inflammatory bowel disease).

# Further Reading

Ravits JM. AAEM minimonograph #48: autonomic nervous system testing. Muscle Nerve. 1997;20(8):919–37.

Freeman R. Autonomic peripheral neuropathy. Lancet. 2005;365(9466):1259–70.

Low PA. Laboratory evaluation of autonomic function. Suppl Clin Neurophysiol. 2004;57:358–68.

.

# Chapter 19
## Neuromuscular Transmission Disorders

## Neurotransmission

Neuromuscular transmission occurs when a quantum of acetylcholine (Ach) from the nerve ending is released and binds to the nicotinic Ach receptors on the postjunctional muscle membrane. The nicotinic Ach receptors on the end plate respond by opening channels to the influx of sodium ions, and subsequent end-plate depolarization leads to muscle contraction. The Ach immediately detaches from the receptor and is hydrolyzed by the enzyme acetylcholinesterase in the synaptic cleft.

## Repetitive Nerve Stimulation

A reliable way of detecting blocking (see definition below) at the neuromuscular junctions in many muscle fibers is by recording, via surface electrodes over the end-plate zone of the entire muscle, while repetitively stimulating the appropriate nerve innervating the muscle. In a healthy muscle, each successive muscle response is the same. In myasthenia gravis (MG), at rest, the size of the initial compound muscle action potential (CMAP) recorded is normal or near normal.

With stimulation at a low frequency of 3–5 Hz, the first two to five stimuli show a decremental response over the weak muscle. This initial decrement is due to progressive

A.Q. Rana et al., *Neurophysiology in Clinical Practice*,
In Clinical Practice, DOI 10.1007/978-3-319-39342-1_19,
© Springer International Publishing Switzerland 2017

blocking at the neuromuscular junction of hundreds of muscle fibers as Ach is depleted, causing an increasing number of end-plate potentials to become subthreshold. This initial fall in the CMAP then levels off, or even slowly increases, during subsequent stimuli. This increase is usually not more than 10–20 %. A reproducible CMAP decrement of 10 % or more between the first and the smallest of the first five responses is considered abnormal. This decrement in the amplitude is considered more characteristic of myasthenic muscles, but a progressive increase in latency may also occur in some of these muscles. To avoid false-negative results in patients with selective involvement of muscles, it is necessary to sample muscles both proximally and distally. Positive tests in the proximal or facial muscles are not uncommon, even when the distal limb muscles reveal no abnormalities.

Rapid neuromuscular stimulation can be achieved by stimulating the nerve at rates between 20 and 40 Hz (painful). The same effect can also be achieved by voluntary exercise of the muscle for several seconds (less painful). Immediately after such an exercise, the size and area of the evoked response may be larger, and the decrement is less. This is called post-activation potentiation (or post-activation facilitation). By assessing the area under the evoked response before and after exercise, post-activation potentiation can be differentiated from "pseudofacilitation." The amplitude of the response is greater after exercise because of the increased synchronization of the components, but the area of the response remains the same.

A few minutes after exercise, the size of the evoked response in the myasthenic muscle may be smaller, and the decrement is greater than at rest. This is called post-activation decrement. This post-activation decrement is helpful in the diagnosis of mildly involved myasthenic muscles, as it may be the only abnormality to occur during repetitive testing. In Lambert-Eaton myasthenic syndromes (pre-synaptic transmission defect), the initial CMAP is low in amplitude, unlike the case in MG, where it is in the normal range. At a low rate of stimulation, a further decrement in amplitude.

In myasthenic syndromes (pre-synaptic transmission defect), the initial CMAP is usually low in amplitude, unlike the case in MG, where it is usually in the normal range. At a low rate of stimulation, a further decrement in amplitude occurs, resembling that in MG. At high rates of stimulation of over 10 Hz, a marked increase in amplitude occurs, and the amplitude can increase to more than 200 % of the initial value.

# Protocol for Repetitive Stimulation in the Diagnosis of Neuromuscular Transmission Disorders

A symptomatic and clinically weak muscle is selected. This could be a facial, proximal, or distal muscle. The electrodes are placed as for a motor nerve conduction study. Movement of the body part is controlled.

1. Establish supramaximal stimulation to 25 % above that producing the maximal CMAP.
2. Stimulate at 2–3 Hz for at least five stimulations.
3. Document any decrement of CMAP by comparing the first response with the fourth or fifth response.
4. Have the patient perform maximum isometric exercise of the muscle for 15 s. If the patient is unable to exercise, then stimulate the muscle at 20 Hz for 10 s.
5. Immediately repeat stimulation at 2–3 Hz for at least five stimulations.
6. Note any increment or decrement in CMAP. A decrement would suggest MG, whereas an increased amplitude of the first response would suggest Lambert-Eaton myasthenic syndrome (LEMS).
7. Wait for 1 min to rest the muscle.
8. Repeat stimulation at 2–3 Hz for at least five stimulations, noting any increment or decrement.

9. Wait for 1 min to rest the muscle.
10. Stimulate at 2–3 Hz for at least five stimulations to note any CMAP decrement or increment. The decrement will be greater in MG, but in other disease conditions the decrement will tend to be stable. A progressive increment of 200 % or more would suggest LEMS, whereas a smaller increment of 50–100 % would suggest botulism.

# Needle Electrode Examination (NEE)

NEE may show variation in the shape and amplitude of the action potential of a single motor unit during a weak voluntary contraction. This variation is a clue to the presence of a neuromuscular transmission defect.

# Single-Fiber Electromyography

Single-fiber electromyography (SFEMG) is a method of monitoring the action potentials of single muscle fibers extracellularly. Recordings may be undertaken with concentric needles, Teflon-coated monopolar needles, or specially designed SFEMG needles. The recording surface should be 25 μm in diameter or smaller. Technical considerations are to filter out lower frequencies, below 1000 Hz. The required characteristics of the EMG machine are to provide a time base with a resolution of a few microseconds, a stable trigger, a delay line, and a method of recording and counting the potentials obtained.

Single-fiber electromyography (SFEMG) is undertaken with a needle electrode that has a very small recording surface, of 25 μm in diameter or less (as stated above), to provide focal recording from only one or two muscle fibers of a motor unit firing during voluntary activity.

A specially designed needle for SFEMG records potentials from one or more muscle fibers of a motor unit that lie within a very small area. The cannula of the needle serves as a reference electrode. When one potential of a pair of fibers suffers total transmission failure this is interpreted as blocking. SFEMG can also be performed by stimulating a motor axon and recording from single muscle fibers.

Fiber density is the mean number of fibers belonging to the same motor unit detected by SFEMG electrodes at a number of different insertion sites. Changes in fiber density reflect reorganization of the motor unit. This alteration in motor unit architecture may be produced by neuropathic as well as myopathic conditions, and may cause normal jitter (see definition below) and blocking. Neuromuscular transmission disorders generally do not alter the motor unit architecture. Normal routine needle EMG should be documented in at least one clinically involved muscle before attributing a pathological jitter or blocking to a neuromuscular transmission disorder.

Standardization of jitter measurements has been most extensive in the extensor digitorum muscle of the forearm.

Commonly, 20 pairs of single fiber action potentials are evaluated, and neuromuscular transmission is considered abnormal if more than 1 of 20 potential pairs have a calculated mean consecutive difference (MCD) of greater than 55 μs or if the mean MCD of the 20 pairs is greater than 35 μs.

# Fiber Density

Single fiber action potentials are obtained by a mild voluntary activation of the muscle in which the recording needle has been placed. The single fiber action potentials should have a duration of approximately 1 ms and a peak-to-peak rise time of 100–200 μs. Also, an amplitude of between 1 and 5 mV electrical activity from only one muscle fiber is recorded in about two-thirds of random needle insertions; this is a reflection of the EMG fiber density of the muscle. The fiber density can be a reproducible measure of the average number of single fiber action potentials in that muscle within the recording radius of the electrode. In neuromuscular disorders, this density is usually normal or only slightly increased. It is greatly increased in reinnervation or in muscular dystrophies.

## Jitter

When the action potentials of two fibers in the same motor unit are recorded simultaneously, the difference in timing of the firing of the two fibers varies. This is called jitter. Jitter is a sensitive measure of the variation in time required for the end-plate potentials of the two fibers to reach threshold and trigger an action potential. Jitter increases when end-plate potentials are low in amplitude. A muscle fiber may intermittently fail to fire when the end-plate potential falls below the threshold for triggering an action potential in the fiber. This is called blocking.

Any given pair of muscle fibers belonging to the same motor unit normally fires in a time-locked manner with only slight variability in the jitter of the interspike intervals, related to physiological and technical factors. Normal jitter varies from muscle to muscle but is generally in the range of 10–50 ms. Jitter in excess of normal, as judged by appropriate statistical analysis after recording from a number of fiber pairs, may occur in a variety of conditions, including neuromuscular transmission disorders.

Pairs of potentials will be recorded approximately 30 % of the time in a normal muscle. The needle records from two muscle fibers sharing the same axon, which is part of the motor unit. The interpotential interval is the time interval between the two potentials. This interval varies from discharge to discharge, and this variability is called jitter. Hence, jitter is produced by fluctuations in the time that end-plate potentials take to reach the threshold for action potential propagation. Jitter is expressed as the MCD to compensate for any slow change in the mean interpotential interval with time. Normal jitter values range between 10 and 60 μs and the value is increased in any condition in which neuromuscular transmission safety factors are lessened and the size and rise time of the end-plate potential are decreased. An increased amount of jitter on SFEMG is a nonspecific measure of neuromuscular dysfunction.

Standardization of jitter measurements has been most extensive in the extensor digitorum muscle of the forearm.

# Blocking

In neuromuscular disorders, if the jitter exceeds by 80–100 μs one of the single fiber potential of a pair may intermittently fail to appear along with the recorded unchanged discharge of the other potential. Hence, blocking occurs when the end-plate potential fails to reach the subthreshold for propagation of the action potential. This blocking is the SFEMG manifestation of clinical fatigue and weakness. This is also the basis for the decremental response that can be seen on a repetitive nerve stimulation study. Increased jitter represents the most sensitive clinical indicator neuromuscular junction, although it remains nonspecific.

Commonly, 20 pairs of single fiber action potentials are evaluated, and neuromuscular transmission is considered abnormal if more than 1 of 20 potential pairs have a calculated MCD of greater than 55 μs or if the mean MCD of the 20 pairs is greater than 35 μs.

# Sensitivity of Various Tests

SFEMG is the most sensitive test for MG (abnormal in 92 % of cases), followed by repetitive nerve stimulation (abnormal in 77 %), and then the Ach receptor (R) antibody assay (abnormal in 73 %). In all cases of MG one of these tests is abnormal.

# Case Study

A 40-year-old woman was referred for evaluation of a 6-month history of difficulty in chewing gum, as well as droopy eyelids, particularly towards the end of the day. Bowel and bladder function was normal and there was no family history of weakness. Physical examination showed slight bilateral ptosis and generalized weakness, but no muscle group atrophy. Sensations and reflexes were normal. A slow repetitive stimulation of the left spinal accessory nerve with recording from trapezius are shown below. Figure 19.1 is a baseline study with no significant decrement, (Fig. 19.2) shows

more than 10% decrement (arrow) one minute after exercise
and (Fig. 19.3) shows more than 20% decrement three minutes after exercise (arrow). She was eventually diagnosed
with myasthenia gravis. Her acetylcholine receptor antibody
titers were three times the upper limit.

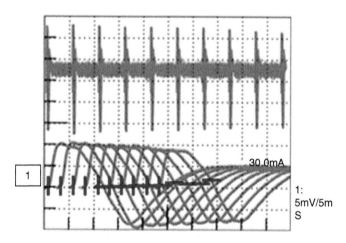

FIGURE 19.1  Baseline study

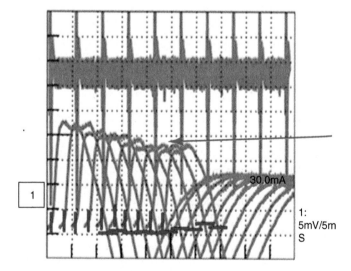

FIGURE 19.2  One minute after exercise (*arrow* note the decrement)

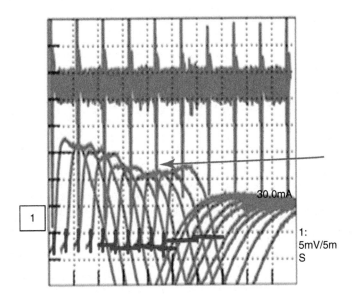

FIGURE 19.3 Three minute after exercise (*arrow* note the decrement)

## Classification of Neuromuscular Transmission Diseases

Neuromuscular transmission diseases are divided into several categories, including autoimmune MG, which presents as adult and juvenile MG, or as transient neonatal myasthenia. Another category is congenital MG, which differs from autoimmune MG because the disrupted communication is not caused by antibodies, but by genetic defects. Myasthenic syndrome (Lambert-Eaton syndrome), with or without carcinoma, is another category, and infantile botulism and botulism in food or wound poisoning form another category. There is also a drug-induced category, with causative drugs that include D-penicillamine, aminoglycosides, cyclosporine, tetracycline, clindamycin, lithium, and chlorpromazine. Lastly, there are a few conditions that do not belong to any of the above categories; these are motor neuron disease, black widow spider poisoning, tick paralysis, neuropathies, myopathies, and myotonias.

## Myasthenia Gravis

Myasthenia gravis (MG) is an autoimmune disorder with antibodies directed against either the acetylcholine receptor (AchR), in 80 % of cases, or postsynaptic muscle-specific kinase (MuSK), in less than 10 % of cases.

## Neonatal Myasthenia Gravis

Neonatal myasthenia gravis (NMG) is an autoimmune disorder that affects one in eight children born to mothers who have been diagnosed with MG. NMG can be transferred from the mother to the fetus by the transmission of AchR antibodies through the placenta. The signs of this disease start at birth, and they include weakness, which responds to anticholinesterase medications. This type of the disease is transient, lasting for about 3 months. However, in some cases, NMG can lead to other health effects, such as arthrogryposis, and even fetal death.

## Lambert Eaton Myasthenic Syndrome

Lambert Eaton myasthenic syndrome (LEMS) is considered to be an autoimmune disorder that affects the presynaptic portion of the neuromuscular junction. LEMS has a unique triad of symptoms, represented by proximal muscle weakness, autonomic dysfunction, and areflexia. Proximal muscle weakness is a product of autoantibodies that are directed against voltage-gated calcium channels, resulting in the reduction of Ach release from motor nerve terminals on presynaptic cells.

## Congenital Myasthenic Syndromes

Congenital myasthenic syndromes (CMS) are almost identical to MG and LEMS in their effects. However, the primary difference between CMS and these two diseases is the genetic origin of CMS. Specifically, CMS are diseases that are typically incurred because of recessive mutations in one of at least ten genes that affect the presynaptic, synaptic, and postsynaptic proteins found in the neuromuscular junction. These syndromes can present at different times throughout the life of an individual. They may arise throughout the fetal phase, causing fetal akinesia, or they may arise in the perinatal period. During these periods, certain conditions, such as ophthalmoplegia, arthrogryposis, ptosis, hypotonia, and feeding or breathing difficulties may be seen. CMS can also present during adolescence or adult years, and cause the individual to develop the slow-channel syndrome.

## Neuromyotonia

Neuromyotonia (NMT), also known as Isaac's syndrome, is a type of neuromuscular disorder caused by hyperexcitability and continuous firing of the peripheral nerve axons. Symptoms include muscle stiffness, myokymia, cramping, increased sweating, and delayed muscle relaxation. These can occur even during sleep or when individuals are under general anesthesia. While they typically occur in the limb and truncal musculature, rare cases also involve the laryngeal and pharyngeal muscles. Many cases are autoimmune from antibodies to voltage gated potassium channels.

# Further Reading

Howard JF. Electrodiagnosis of disorders of neuromuscular transmission. Phys Med Rehabil Clin N Am. 2013;24:169–92.

Sanders DB, Howard JF. AAEE minimonograph #25 single fiber electromyography in myasthenia gravis. Muscle Nerve. 1986;9(9):809–919. Article first published online: 13 Oct 2004.

Vincent A, Bowen J, Newsome-Davis J, McConville J. Seronegative generalised myasthenia gravis: clinical features, antibodies and their targets. Lancet Neurol. 2003;2(2):99–106.

Oh SJ, Kim DE, Kuruoglu R, et al. Diagnostic sensitivity of the laboratory tests in myasthenia gravis. Muscle Nerve. 1992;15:720–4.

.

# Chapter 20
# Myopathy

Electromyography (EMG) is a procedure for screening patients with myopathies, and it remains the most common and the most important technique for assessing the course of these diseases over time. For a definitive diagnosis, molecular genetics and muscle biopsy are required. Fibrillation potentials, myotonic or complex repetitive discharges, positive sharp waves, and polyphasic potentials are nonspecific and can occur in both myopathic and neurogenic lesions.

The most sensitive and specific parameter for myopathy in conventional electrodiagnosis is the EMG pattern generally referred to as "myopathic." This is one of brief duration, small amplitude, and abundant recruitment of motor unit action potentials upon slight voluntary effort. However, this pattern can also be observed in disorders of the neuromuscular junction or the terminal motor fibers. More advanced techniques, such as macro EMG, single-fiber EMG, scanning EMG, and turns/amplitude analysis have revealed additional possibilities for analysis of the motor unit and the interference pattern see Table 20.1 for differences between neuropathic and myopathic potentials.

A.Q. Rana et al., *Neurophysiology in Clinical Practice*,
In Clinical Practice, DOI 10.1007/978-3-319-39342-1_20,
© Springer International Publishing Switzerland 2017

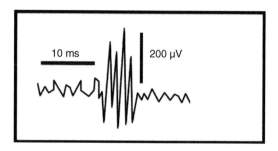

FIGURE 20.1  Myopathic motor unit potential

TABLE 20.1  Electromyographic findings in various conditions

| Abnormality | Fibrillation | Polyphasic motor unit potential (MUP) | Brief small-amplitude polyphasic MUP |
|---|---|---|---|
| Myopathy | + | + | + |
| Acute denervation | + | – | – |
| Chronic denervation | ± | + | – |
| Neuromuscular junction disorders | ± | – | ± |

## Myopathies with Myotonia

Myotonia is described clinically as the occurrence of delayed relaxation of a muscle after a voluntary contraction or percussion. Myotonia is often easier to recognize on EMG examination than on neurological examination. Myotonic potentials are caused by chronically depolarized muscle membranes and are noted as spontaneous, painless discharges, with waxing and waning of both amplitude and frequency producing a characteristic audio profile, described as sounding like a dive-bomber.

Myotonic potentials are repetitive discharges at a rate of 20–80 Hz. They can be biphasic spike potentials less than

5 ms in duration, or they can be described as positive waves of 5–20 ms in duration that resemble positive sharp waves. A single myotonic potential would look and sound just like a fibrillation potential or a positive sharp wave. Needle insertion and movement, muscle contraction, or tapping the muscle will bring about myotonia.

# Differential Diagnosis of Myotonic Disorders

1. Clinical myotonia and electrical myotonia:

   - Myotonic dystrophy type I (DM1)
   - Myotonic dystrophy type II (DM2; proximal myotonic myopathy)
   - Myotonia congenita (e.g., Thomsen disease)
   - Schwartz-Jampel syndrome

2. Clinical paramyotonia and electrical myotonia:

   - Hyperkalemic periodic paralysis
   - Paramyotonia congenita

3. Electric myotonia without clinical myotonia:

   - Acid maltase deficiency

4. Uncommon causes of myotonia:

   - Myopathy (centronuclear/myotubular)
   - Drug-induced (e.g., by chloroquine, colchicine)

# Electrodiagnostic Studies that Can Aid Differential Diagnosis

1. Repetitive stimulation. In myotonic syndromes, repetitive stimulation at 5–10 Hz leads to a decrement in the compound muscle action potential (CMAP).
2. Exercise testing. After short exercise of 10–30 s, the CMAP is recorded and compared with the one recorded prior to exercise. In patients with DM1, this brief period of exercise

causes a decrease in the CMAP, whereas in DM2 there is no change.

3. Cooling test. Cold provokes the symptoms of weakness in patients with paramyotonia congenita. This feature can be reproduced in the electrodiagnostic laboratory by recording the CMAP of an individual muscle before and after cooling of the limb for 15–30 min at 15 °C. The CMAP decrement in paramyotonia congenita patients is typically greater than 75%.

# Myopathies with Denervating Features

*Dystrophies*

- Dystrophin deficiency – Duchenne and Becker
- Facioscapulohumeral muscular dystrophy
- Autosomal recessive distal muscular dystrophy
- Emery-Dreifuss muscular dystrophy
- Oculopharyngeal muscular dystrophy

## *Inflammatory Myopathies*

- Polymyositis
- Dermatomyositis
- Inclusion body myositis
- HIV-associated myositis

## *Congenital Myopathies*

- Centronuclear/myotubular myopathy
- Nemaline rod myopathy

## *Metabolic Myopathies*

- Acid maltase deficiency myopathy
- Carnitine deficiency myopathy
- Debrancher deficiency myopathy

## *Drug-induced and Toxic Myopathies*

- Steroid myopathy
- Colchicine, alcohol, chloroquine, and clofibrate myopathies

- Statin myopathy
- Critical illness myopathy

*Infectious Myopathies*

- Trichinosis
- Toxoplasmosis

*Amyloid Myopathy:* This is a rare presentation of primary systemic amyloidosis (AL). Like inflammatory myopathies, it can present with proximal muscle weakness and an increased creatine kinase level. The usual symptoms of are usually nonspecific, including progressive proximal weakness with an elevated creatine kinase level, in addition patients can have macroglossia and muscle pseudohypertrophy.

*Autoimmune necrotizing myopathy:* Autoimmune necrotizing myopathy is a rare form of idiopathic inflammatory myopathy characterized clinically by acute or subacute proximal muscle weakness, and histopathologically by myocyte necrosis and regeneration without significant inflammation. The presenting feature is subacute severe symmetrical proximal myopathy, with a markedly elevated creatine kinase level. The course is often severe but may be self-limiting in certain cases. The disease is thought to be related to an immune response possibly triggered by drug therapy (statins), connective tissue diseases, or cancer.

*Recommended Electromyographic Approach to Myopathy*
*Routine Studies*

1. At least two distal and two proximal muscles in the lower extremity
2. At least two distal and two proximal muscles in the upper extremity
3. At least one paraspinal muscle

*Special Considerations*

- Always try to study the symptomatic and weak muscle.
- The muscle on the contralateral side (deltoid, biceps, vastus lateralis, and gastrocnemius) of the selected muscle should be easily biopsied.

- Consider non-routine evaluation methods, such as:

  (a) Quantitative motor unit action potential (MUAP) analysis: accumulate 20 MUAPs from different locations within each muscle. Calculate the mean amplitude and duration and compare with findings in muscle samples from age-matched control.
  (b) Single-fiber EMG: repetitive nerve stimulation studies can also be performed.

# Case Study

A 45-year-old woman attends the neuromuscular clinic for progressive weakness of 10-month duration. She has a long-standing history of bronchial asthma, treated with prednisone. Initial symptoms were those of weakness, particularly in climbing stairs and getting out of chairs. She has also developed dysphagia. There is no history of pain or paresthesia.

Neurological examination shows Medical Research Council grade 4/5 proximal muscle weakness in all limbs, with difficulty in rising from a squat and difficulty in getting out of a chair unassisted. There is also mild weakness of neck flexion, but neck extension is normal. The rest of the neurological examination is normal.

EMG findings are abnormal, with diffuse fibrillation potentials in the proximal muscles, and many of the MUAPs are of brief duration, polyphasic, and low amplitude, with an early recruitment pattern. The differential diagnosis here is of steroid myopathy versus polymyositis.

## *Duchenne Muscular Dystrophy*

Duchenne muscular dystrophy is an X-linked genetic disorder that results in the absence of a structural protein, known as dystrophin, at the neuromuscular junction. This disease affects 1 in 3600–6000 males, and it frequently causes death by the age of 30. The absence of this structural protein, dystrophin, causes muscle degeneration. Patients usually present with abnormal gait, hypertrophy in the calf muscles, and an elevated CK level.

# Further Reading

Fuglsang-Frederiksen A. The role of different EMG methods in evaluating myopathy (invited review). Clin Neurophysiol. 2006;117:1173–89.

Heatwole CR, Statland JM, Logigian EL. The diagnosis and treatment of myotonic disorders. Invited reviews. Muscle Nerve. 2013;47:632–48.

TM. Differential diagnosis of myotonic disorder in AANEM monograph #27, AANEM March 2008.

# Chapter 21
## Long Latency Reflexes

### F-Wave

The F-wave is an action potential that is evoked intermittently from a muscle by a supramaximal electrical stimulation of the nerve that is caused by the antidromic activation of motor neurons. The F-wave action potential is approximately 1–5 % that of the M-wave and has a variable configuration. Its latency is longer than that of the M-wave and is variable

F-wave studies evaluate the conduction of motor axons proximal to the stimulation site. During routine studies, the antidromic potential reaches and depolarizes the cell body. The depolarization is conducted back to the axon, and a new action potential is generated, which conducts to the muscle. This response is the F-wave.

The recording electrodes for F-waves are placed in the same location as those for motor nerve conduction studies. The stimulating electrodes are reversed so that the cathode is towards the spine.

Any demyelinating peripheral neuropathy can cause the slowing of F-wave latency. Abnormal F-wave responses are the earliest finding in Guillain-Barré syndrome. F-waves may be unrecordable in severe demyelinating neuropathies, but F-wave latency is usually normal in axonopathy, radiculopathies, and plexopathies, except in severe axonopathy, when secondary demyelination may occur.

A.Q. Rana et al., *Neurophysiology in Clinical Practice*,
In Clinical Practice, DOI 10.1007/978-3-319-39342-1_21,
© Springer International Publishing Switzerland 2017

# H-Wave

The H-wave is a compound muscle action potential (CMAP) with a consistent latency recorded from muscles after stimulation of the nerve; it has a longer latency than the M-wave of the same muscle. The H-wave is most reliably elicited from the gastroc/soleus complex. Stimulation of long duration (500–100 ms) – the stimulus intensity sufficient to elicit a maximal amplitude M-wave – reduces or abolishes the H-wave. This reduction or abolition is thought to be due to a spinal reflex, with electrical stimulation of afferent fibers in a mixed nerve and activation of motor neurons to the muscle mainly occurring through a monosynaptic connection in the spinal cord. The latency of the H-wave is longer at more distal sites of stimulation.

The H-reflex is the electrophysiological counterpart of the ankle jerk. The normal reflex is 35 ms, and the difference between the two sides should be less than 1.4 ms. For recording H-reflexes, the tibial nerve is stimulated. The stimulating electrode is placed in the popliteal fossa and recording electrodes are placed on the gastrocnemius muscle. A submaximal stimulus is given. The first visible response at about 30 ms is the H-reflex. The normal latency of the H-reflex is 35 ms. A delayed or absent H-reflex is associated with demyelinating and axonal neuropathies, and this feature may be seen in S1 radiculopathy as well.

# A-Wave

The use of the term "axon reflex" instead of "A-wave" is discouraged, as no reflex is involved. The A-wave is a CMAP that follows the M-wave; the A-wave is evoked consistently from a muscle by submaximal stimulation and is frequently abolished by supramaximal stimulation. Its amplitude is similar to that of the F-wave, but its latency is constant. The A-wave usually occurs before the F-wave, but it may occur after the F-wave. It is thought to be caused by extra discharges in nerve branchings or ephapses.

# Chapter 22
# Cranial Nerve Studies

Facial neuropathy is the most common cranial neuropathy. The anatomy of the facial nerve is complex, with a long intracranial course. Electrodiagnosis can be helpful in determining prognosis, but such studies are not helpful until several days after the onset of the neuropathy. The clinical presentation shows unilateral weakness of the upper and lower parts of the face, hyperacusis, dysgeusia, and abnormal lacrimation and salivation. Seventy percent of cases are idiopathic, or they can be regarded as Bell's palsy. Ramsay Hunt syndrome, also known as herpes zoster oticus, is another common cause of facial neuropathy.

The electrodiagnostic methods commonly used to study the facial nerve are direct facial motor nerve conduction studies, needle electrode electromyography of facial-innervated muscles, and the blink reflex. Electrodiagnostic studies are quite useful for the determination of demyelinating pathology or axonopathy. Wallerian degeneration of motor fibers takes 5–8 days after an axonal injury, and assessment of prognosis is possible only after this period.

## Value of Compound Muscle Action Potential

The amplitude of the compound muscle action potential (CMAP), with side-to-side comparison, determined 5–7 days after disease onset, has been used to assess prognosis. When the CMAP amplitude is less than 10 % of that on the healthy side, maximum recovery will take 6–12 months, and function will be moderately or severely limited. If the amplitude is 10–30 % that of the healthy side, recovery may take 2–8 months, with mild to moderate residual weakness. If the CMAP amplitude is >30 % that of the healthy side, complete recovery can be expected by 2 months after disease onset.

## Value of Latency

The latency of direct facial motor nerve stimulation is not as useful as CMAP for assessing prognosis. In a latency study done 5–7 days after disease onset, normal latency assures patients of a complete recovery, without aberrant features. With prolonged latency compared with that of the opposite side, a good recovery is probable, but with some chance of synkinesis. With no latency response, there is a high incidence of synkinesis, no recovery, or both.

## Needle Electrode Electromyography

Needle electrode electromyography can potentially detect abnormalities in facial-innervated muscles, within hours of disease onset, in the form of a decreased interference pattern or increased motor unit potential firing rate. The prognostic value is low, as these abnormalities do not help to differentiate demyelinating from axonal lesions. However, the existence of even a few volitional motor unit potentials (MUPs) in a patient with full clinical paralysis indicates the nerve is still in continuity and is consistent with a favorable prognosis.

Fibrillations and positive waves indicate the presence of axonal degeneration, with probably a prolonged and incomplete recovery. These features are unlikely to be seen earlier than 1–2 weeks after disease onset and may not be observed for up to 3 weeks. Electromyography (EMG) can also be helpful in assessing prognosis, as volitional MUPs reappear at a time when the movement of facial muscles is not yet visible. Unstable polyphasic MUPs imply ongoing reinnervation and suggest a favorable recovery.

# The Blink Reflex

The blink reflex is the electrical correlate of the bedside corneal reflex. It differs from a direct facial nerve study in that determination of this reflex involves examination of the trigeminal nerve and the pons, as well as the facial nerve. The blink reflex can be used to assess proximal segments of the facial nerve that are inaccessible to the direct stimulation technique. An abnormal blink reflex showing a pattern consistent with facial neuropathy has been found in a variety of conditions, such as Bell's palsy, inflammatory demyelinating neuropathy, hereditary neuropathy, diabetes, multiple sclerosis, and acoustic neuroma.

Facial motor synkinesis can be assessed by expanding the set-up for the blink reflex to also include recording electrodes over the orbicularis oris or other facial muscles. In a healthy patient, supraorbital nerve stimulation results in a response recorded only from the orbicularis oculi. In patients with aberrant regeneration, a response will be found in both the orbicularis oculi and the orbicularis oris. The blink reflex has not been particularly helpful as a prognostic method, as it offers little beyond what is offered by direct facial nerve studies. Also, it is limited by the same time constraints as the direct studies.

# Blink Reflex Studies

Blink reflex studies measure the entire reflex arch between the trigeminal and the facial nerves, including the proximal segments of the facial nerves. Hence, to evaluate the proximal facial nerve segments, a blink reflex study is used in combination with facial nerve stimulation studies.

Lesions of the facial nerve result in abnormalities of the ipsilateral R1 and R2 components of the blink reflex, whereas the contralateral R2 response remains normal. When the normal contralateral side is stimulated, the opposite pattern is seen; normal ipsilateral R1 and R2 responses and an abnormal contralateral R2 response. The table (22.1) below provides a brief overview of blink reflex interpretation.

TABLE 22.1  Blink reflex and its interpretation

| Lesion | Affected side | Unaffected side |
|---|---|---|
| Trigeminal neuropathy | Delay/absent ipsilateral R1/R2 and contralateral R2 | Normal ipsilateral R1/R2 and contralateral R2 |
| Facial neuropathy | Delay/absent ipsilateral R1/R2 with normal contralateral R2 | Normal ipsilateral R1/R2 but prolonged/absent contralateral R2 |
| Unilateral pontine lesion | Delay/absent ipsilateral R1 but normal ipsilateral/contralateral R2 | Normal ipsilateral R1/R2 and contralateral R2 |
| Unilateral medullary lesion | Normal ipsilateral R1 and contralateral R2 with delay/absent ipsilateral R2 | Normal ipsilateral R1/R2 delay/absent contralateral R2 |
| Demyelinating neuropathies | Delay/absent ipsilateral R1/R2 and contralateral R2 | Delay/absent ipsilateral R1/R2 and contralateral R2 |
| Axonal neuropathies | Normal ipsilateral R1/R2 and contralateral R2 | Normal ipsilateral R1/R2 and contralateral R2 |

# Electrophysiological Evaluation of Facial and Trigeminal Nerve Lesions

## Facial Nerve

1. Facial nerve studies:

    (a) Stimulation of the facial nerve trunk can be performed below and anterior to the mastoid or directly anterior to the tragus, with recording from a facial muscle (nasalis and orbicularis oculi).

    (b) Individual stimulation of facial branches can be undertaken, e.g., stimulation of the following branches:

- Frontal branch
- Zygomatic branch
- Mandibular branch

2. Blink reflex studies, undertaken by stimulating the supra-orbital nerve and recording from the orbicularis oculi muscles, using bilateral studies
3. Needle electromyographic examination of the frontalis, orbicularis oculi, orbicularis oris, or the mentalis muscles

## *Trigeminal Nerve*

1. Blink reflex studies, undertaken by stimulating the supra-orbital nerve and recording from the orbicularis oculi muscles bilaterally
2. Needle electromyographic examination of the masseter and temporalis muscles

# Chapter 23
## Somatosensory Evoked Potentials

Somatosensory evoked potentials (SEPs) are used as an extension of the electrodiagnostic evaluation and nerve conduction tests that are performed in large myelinated sensory fibers of the peripheral and central nervous systems. SEP studies are noninvasive; SEPs are obtained by the repetitive submaximal stimulation of a sensory or mixed sensory/motor peripheral nerve and recording the average responses from electrodes placed over proximal portions of the nerves stimulated in the plexus, spine, and scalp.

Current clinical applications of SEPs assess short latency responses. These are portions of a SEP waveform occurring within 25 ms after stimulation of a nerve at the wrist, 40 ms after stimulation of a nerve at the knee, and 50 ms after stimulation of a nerve at the ankle. The amplitude, peak, and interpeak latency measurements, with side-to-side comparisons, are used to assess abnormalities. Depending on the clinical condition being studied, several nerves in one extremity may have to be tested and the results compared with those for the contralateral limb to assess focal and diffuse disease processes.

SEPs may also be used to assess the functional integrity of central and peripheral sensory pathways. Common conditions in which SEPs have demonstrated usefulness in electrodiagnostic medicine include, but are not limited to, the following: mononeuropathy, polyneuropathy, plexopathy,

A.Q. Rana et al., *Neurophysiology in Clinical Practice*,
In Clinical Practice, DOI 10.1007/978-3-319-39342-1_23,
© Springer International Publishing Switzerland 2017

neuronopathy, radiculopathy, spinal cord trauma, subacute combined degeneration, non-traumatic spinal cord lesions, multiple sclerosis (MS), spinal cerebellar degeneration, myoclonus, coma, and the intraoperative monitoring of peripheral/cranial nerves, the spinal cord, and the brain stem/brain.

SEPs may be helpful in studying disorders of the brain, brain stem, spinal cord, dorsal roots, and peripheral nerves. SEPs are often helpful in localizing the anatomical site of a lesion in the somatosensory pathway. SEPs may be used to identify conduction abnormalities caused by axonal loss, demyelination, or both. SEP abnormalities are not disease-specific, but can indicate conduction impairments associated with certain disorders. Also, SEPs are useful in confirming nonorganic sensory loss. In such cases, SEPs generated from the stimulation of virtually any numb area may be compared with recordings obtained from stimulation of the asymptomatic contralateral area.

# Brain and Brain Stem

SEP abnormalities reflecting pathology in the brain or the spinal cord are found in up to 90 % of patients with diagnosed MS and in approximately 50 % of MS patients who have no current sensory signs or symptoms.

Lower limb SEPs are more likely to be abnormal than those of the upper limb in MS. However, both upper and lower limb SEP testing are often indicated, because the patients may demonstrate abnormalities in only one of these regions.

SEPs may also be abnormal in other diseases that affect myelin; in hereditary nervous system degeneration, such as Friedreich's ataxia; in myoclonus; and in coma; there are also intraoperative indications for the use of SEP testing.

# Spinal Cord

Abnormal SEP findings may be recorded over the spinal cord in many disorders that affect the ascending pathways, such as MS. SEPs are useful in the evaluation of spinal cord trauma, subacute combined degeneration, cervical spondylosis and myelopathy, syringomyelia, hereditary spastic paraplegia, transverse myelitis, MS, vascular lesions, spinal cord tumors, myelomeningocele, and tethered cord syndrome; SEPs are also useful for spinal cord monitoring during surgery.

# Ventral Root Lesions

Dermatomal SEPs have been used to evaluate acute radiculopathies. These SEPs may be useful in conditions such as lumbar stenosis and cervical root diseases.

# Further Reading

The inquisitive reader is directed to several excellent texts that are available for comprehensive understanding of the vast discipline of electrophysiology and electrodiagnostic evaluation of the peripheral nervous system.

Chiappa K. Short latency somatosensory evoked potentials: interpretation. In: Chiappa K, editor. Evoked potentials in clinical medicine. New York: Raven; 1990. p. 400–7.

Dumitru D. Electrodiagnostic medicine. 2nd ed. Philadelphia: Hanley & Belfus, Inc.; 2002.

Katirji B. Electromyography in clinical practice: a case study approach. 2nd ed. Philadelphia: Mosby Elsevier; 2007.

Kimura J. Electrodiagnosis in diseases of nerve and muscles: principles and practice. 4th ed. Philadelphia: FA Davis; 1989.

Krivickas LS, editor. Electrodiagnosis in neuromuscular disorders, Physical Medicine and Rehabilitation Clinics of North America. Philadelphia: WB Saunders Company; 2003.

Oh SJ. Clinical electromyography. Nerve conduction studies. 3rd ed. Philadelphia: Lippincott Williams & Wilkins; 2003.

Preston DC, Shapiro B. Electromyography and neuromuscular disorders: clinical-electrophysiologic correlations. 3rd ed. New York: Elsevier Saunders; 2013.

# Appendix

TABLE I Nerve root supplies of commonly tested muscles

| Upper limb | Spinal roots |
|---|---|
| **Spinal accessory nerve** | |
| Trapezius | C3, C4 |
| | |
| **Brachial plexus** | |
| Rhomboids – Dorsal scapular nerve | C4, C5 |
| Serratus anterior – Long thoracic nerve | C5, C6, C7 |
| Pectoralis major | |
|     Clavicular ⎫ | C5, C6 |
|     Sternal   ⎭ | C6, C7, C8 |
| Supraspinatus | C5, C6 |
| Infraspinatus | C5, C6 |
| Latissimus dorsi – Thoracodorsal | C6, C7, C8 |
| Teres major – Subscapular nerve | C5, C6, C7 |
| | |
| **Axillary nerve** | |
| Deltoid | C5, C6 |
| | |
| **Musculocutaneous nerve** | |
| Biceps | C5, C6 |
| Brachialis | C5, C6 |
| | |
| **Radial Nerve** | |
|           Long head ⎫ | |
| Triceps    Lateral head ⎬ | C6, C7, C8 |
|           Medial head ⎭ | |
| Brachioradialis | C5, C6 |
| Extensor carpi radialis longus – wrist extension | C5, C6 |
| | |
| **Posterior interosseous nerve** | |
| Supinator | C6, C7 |
| Extensor carpi ulnaris | C7, C8 |
| Extensor digitorum | C7, C8 |
| Abductor pollicis longus | C7, C8 |
| Extensor pollicis longus | C7, C8 |

A.Q. Rana et al., *Neurophysiology in Clinical Practice*, In
Clinical Practice, DOI 10.1007/978-3-319-39342-1,
© Springer International Publishing Switzerland 2017

TABLE I (continued)

| Upper limb | Spinal roots |
|---|---|
| Extensor pollicis brevis | C7, C8 |
| Extensor indicis | C7, C8 |
| | |
| *Median nerve* | |
| Pronator teres | C6, C7 |
| Flexor carpi radialis | C6, C7 |
| Flexor digitorum superficialis | C7, T1 |
| Abductor pollicis brevis | C8, T1 |
| Flexor pollicis brevic* | C8, T1 |
| Opponens pollicis | C8, T1 |
| Lumbricals I and II | C8, T1 |
| | |
| *Anterior Interosseous Nerve* | |
| **Pronator quadratus** | C7, C8 |
| **Flexor digitorum profundus I and II** | C7, C8 |
| **Flexor pollicis longus** | C7, C8 |
| | |
| *Ulnar nerve* | |
| **Flexor carpi ulnaris** | C7, C8, T1 |
| **Flexor digitorum profundus III and IV** | C7, C8 |
| **Hypothenar muscle** | C8, T1 |
| **Adductor pollicis** | C8, T1 |
| **Flexor pollicis brevis – also by median** | C8, T1 |
| **Palmar interossei** | C8, T1 |
| **Dorsal interossei** | C8, T1 |
| **Lumbricals III and IV** | C8, T1 |

*Flexor pollicis brevis is often supplied wholly or partially by the ulnar nerve.

| Lower limb | Spinal roots |
|---|---|
| | |
| *Femoral nerve* | |
| **Illiopsoas** | L1, L2, L3 |
| **Rectus femoris** | |
| **Vastus lateralis**  } **Quadriceps** | L2, L3, L4 |
| **Vastus intermedius femoris** | |
| **Vastus medialis** | |
| | |
| *Obturator nerve* | |
| **Adductor longus** } | L2, L3, L4 |
| **Adductor magnus** | |
| | |
| *Superior gluteal nerve* | |
| **Gluteus medus and minimus** } **Hip abduction** | L4, L5, S1 |
| **Tensor fasciae latae** | |
| | |

TABLE I (continued)

| Lower limb | Spinal roots |
|---|---|
| *Inferior gluteal nerve* | |
| Gluteus maximus | L5, S1, S2 |
| | |
| *Sciatic and tibial nerve* | |
| Semitendinosus | L5, S1, S2 |
| Biceps femoris | L5, S1, S2 |
| Semimembranosus | L5, S1, S2 |
| Gastrocnemius and soleus | S1, S2 |
| Tibialis posterior | L4, L5 |
| Flexor digitorum longus | L5, S1, S2 |
| Abductor hallucis ⎫ | |
| Abductor digiti minimi ⎬ Small muscles of foot | S1, S2 |
| Interossei ⎭ | |
| | |
| *Sciatic and common personeal nerve* | |
| Tibialis anterior | L4, L5 |
| Extensor digitorum longus | L5, S1 |
| Extensor hallucis longus | L5, S1 |
| Extensor digitorum brevis | L5, S1 |
| Peroneus longus | L5, S1 |
| Peroneus brevis | L5, S1 |

This list includes only commonly tested muscles innervated by these nerves, with the order of innervations.

TABLE 2 Muscles and nerves involved in different movements

| Movement | Movements weak in UMN lesions | Root | Reflex | Nerve | Muscle |
|---|---|---|---|---|---|
| *Upper limb* | | | | | |
| Shoulder abduction | ++ | C5 | | Axillary | Deltoid |
| Elbow flexion | | C5/6 | + | Musculocutaneous | Biceps |
| | | C6 | + | Radial | Brachioradialis |
| Elbow extension | + | C7 | + | Radial | Triceps |
| Radial wrist extension | + | C6 | | Radial | Extensor carpi radialis longus |
| Finger extension | + | C7 | | Posterior interosseous nerve | Extensor digitorum communis |
| Finger flexion | | C8 | + | Anterior interosseous nerve | Flexor pollicis longus + flexor digitorum profundus (*index*) I, II |
| | | | | Ulnar | Flexor digitorum profundus 2.4 (ring + little) |
| Finger abduction | ++ | T1 | | Ulnar | First dorsal interosseous |
| | | T1 | | Median | Abductor pollicis brevis |

*Lower limb*

| | | | | |
|---|---|---|---|---|
| Hip flexion | ++ | L1/2 | | Iliopsoas |
| Hip adduction | | L2/3 | Obturator | Adductors |
| Hip extension | + | L5/S1 | Inferior gluteal nerve | Gluteus maximus |
| Knee flexion | | S1 | Sciatic | Hamstring |
| Knee extension | + | L3/4 | Femoral | Quadriceps |
| Ankle dorsiflexion | ++ | L4 | Deep peroneal | Tibialis anterior |
| Ankle eversion | | L5/S1 | Superficial peroneal | Peroneal muscle |
| Ankle plantar flexion | + | S1/S2 | Tibial | Gastrocnemius, soleus |
| Big toe extension | | L5 | Deep peroneal | Extensor hallucis longus |

This table shows the principal muscles with their roots and nerve supplies in which movements are commonly tested
*UMN* upper motor neuron

TABLE 3  Common electromyographic abnormalities

| Finding | Characteristics | Interpretation |
|---------|-----------------|----------------|
| *Resting activity* | | |
| Fibrillations | Single muscle fiber action potentials due to membrane instability | Acute denervation and myopathy |
| Fasciculations | Spontaneous discharge of a motor unit | Denervation and motor neuron disease |
| Positive sharp wave | Single muscle fiber action potential due to membrane instability | Acute denervation and myopathy. May appear before fibrillations |
| Myotonia | Repetitive discharge of muscle fibers | Myotonic dystrophy, myotonia congenital, and sometimes in periodic paralysis |
| Complex repetitive discharge | Repetitive discharge of muscle fibers | Denervation and myopathy |
| *Motor unit activity* | | |
| Brief small-amplitude polyphasic MUPs | Decreased number of functioning muscle fibers | Myopathy |
| Long-duration polyphasic MUPs | Increased number of muscle fibers innervated by an axon | Chronic denervation |
| *Maximal contraction* | | |
| Early recruitment | Seen when more units have to be recruited at faster rates even with low effort | Myopathy |

TABLE 3 (continued)

| Finding | Characteristics | Interpretation |
|---------|-----------------|----------------|
| Reduced recruitment | Seen when reduced numbers of functioning motor units are recruited | Denervation |

*MUP* motor unit potential

TABLE 4 NCS and EMG findings in common disorders

| Disorder | NCV | EMG finding |
|----------|-----|-------------|
| Demyelinating neuropathy | Slowing of motor and sensory NCVs, prolongation of F-waves | Reduced recruitment |
| Axonal neuropathy | Reduction of CMAP amplitude, normal NCVs | Long-duration polyphasic MUP |
| Myopathy | NCVs are normal, may have decreased CMAP amplitude | Small polyphasic MUP |
| Motor neuron degeneration | NCVs are normal, may have reduced CMAP amplitude | Fasciculations, fibrillations |
| Neuromuscular junction disorder | Normal or reduced CMAP amplitude, abnormal repetitive stimulation | Normal |

*EMG* electromyography, *NCS* nerve conduction studies, *NCVs* nerve conduction velocities, *CMAP* compound motor action potential

.

# Index

A.Q. Rana et al., *Neurophysiology in Clinical Practice*,
In Clinical Practice, DOI 10.1007/978-3-319-39342-1,
© Springer International Publishing Switzerland 2017